Modeling and Simulation in Science, Engineering and Technology

Series Editor
Nicola Bellomo
Politecnico Torino
Italy

Advisory Editorial Board

K.J. Bathe
Massachusetts Institute of Technology
USA

W. Kliemann
Iowa State University
USA

S. Nikitin
Arizona State University
USA

V. Protopopescu
CSMD
Oak Ridge National Laboratory
USA

P. Degond
Université P. Sabatier Toulouse 3
France

P. Le Tallec
INRIA
France

K.R. Rajagopal
Texas A&M University
USA

Y. Sone
Kyoto University
Japan

E.S. Subuhi
Istanbul Technical University
Turkey

Gilbert G. Walter
Martha Contreras

Compartmental Modeling with Networks

Birkhäuser
Boston • Basel • Berlin

Gilbert G. Walter
Department of Mathematical Sciences
University of Wisconsin–Milwaukee
P.O. Box 413
Milwaukee, WI 53201-0413
USA

Martha Contreras
Department of Biometry
Cornell University
434 Warren Hall
Ithaca, NY 14853-7801
USA

Library of Congress Cataloging-in-Publication Data
Walter, Gilbert G.
    Compartmental modeling with networks / Gilbert Walter, Martha
Contreras.
        p.   cm. — (Modeling and simulation in science, engineering and
    technology)
    Includes bibliographical references and index.
    ISBN 0-8176-4019-3 (alk. paper)
    1. Mathematical models.   2. Computer simulation.   I. Contreras,
Martha.   II. Title.   III. Series.
TA342.W35   1999
511′.8—dc21
                                                                                            98-44616
                                                                                            CIP

AMS Subject Classifications: 05, 60, 92

Printed on acid-free paper.
© 1999 Birkhäuser Boston   *Birkhäuser* 🅱 ®

ISBN 0-8176-4019-3
ISBN 3-7643-4019-3

Formatted from the authors' TeX files.
Printed and bound by Edwards Brothers, Inc., Ann Arbor, MI.
Printed in the United States of America.

9 8 7 6 5 4 3 2 1

Contents

Preface

The subject of mathematical modeling has expanded considerably in the past twenty years. This is in part due to the appearance of the text by Kemeny and Snell, "Mathematical Models in the Social Sciences," as well as the one by Maki and Thompson, "Mathematical Models and Applications." Courses in the subject became a widespread if not standard part of the undergraduate mathematics curriculum. These courses included various mathematical topics such as Markov chains, differential equations, linear programming, optimization, and probability. However, if our own experience is any guide, they failed to teach mathematical modeling; that is, few students who completed the course were able to carry out the modeling paradigm in all but the simplest cases. They could be taught to solve differential equations or find the equilibrium distribution of a regular Markov chain, but could not, in general, make the transition from "real world" statements to their mathematical formulation. The reason is that this process is very difficult, much more difficult than doing the mathematical analysis. After all, that is exactly what engineers spend a great deal of time learning to do. But they concentrate on very specific problems and rely on previous formulations of similar problems. It is unreasonable to expect students to learn to convert a large variety of real-world problems to mathematical statements, but this is what these courses require.

Fortunately, there is a large class of problems for which the transition step is not as difficult. These are problems for which the appropriate model is a flow model. They are used when there is a flow of something such as a fluid or money or energy between the components of a system. They are widely used in biomedicine for tracer experiments, but have applications in other areas of biology such as the study of ecosystems, as well as in input–output analysis in economics, arms races, and the study of epidemics. What's more is that the transition step from the problem statement to the mathematical formulation is very intuitive and easily learned. Most of our students chose such models for their course projects even though they formerly constituted a small portion of the course.

These flow models are usually called *compartmental models* to distinguish them from diffusion models. To construct them, a system is divided up into homogeneous compartments and the flows of material between the various compartments and to and from the outside are traced. This leads immediately to a directed graph, which can be used as a simple model and may be adequate for an initial analysis of the system. If information on flow rates is known, the model becomes a system of differential or difference equations. These can be solved by traditional means, but because of the special nature of these models, properties of the solutions can be inferred without knowing the solutions explicitly.

In this work, we shall concentrate on these models and their applications. This will involve consideration of some properties of digraphs (directed graphs), their relations to the systems of differential equations, and their conversion to Markov chains. The applications will be to ecosystem models, to fluid transfer, to competition models, to predator-prey systems, to fisheries management, to regular and absorbing Markov chain models, to Leontief input-output models, to Leslie matrices, to tracer experiments, to epidemic models, and to network flows in engineering.

The book will be organized in four parts as follows:

The first part will be devoted to the theory of digraphs; it will be at a level accessible to most students who have had any college mathematics course.

The second part deals with Markov chains. Although the terminology involves concepts from probability theory, the mathematics is pure matrix theory. It requires only the material found in a first course in matrix theory or finite mathematics.

The third part will introduce the differential equations associated with compartmental models and develop their theory to some extent. This part will require a knowledge of calculus and matrices, but not necessarily differential equations.

The final part will go into the theory of compartmental models somewhat more deeply. The relations between the dynamics of the solution and the structure of the model will be studied. The required level of sophistication is higher but still requires only calculus and matrix theory.

Although these compartmental models are not designed to be computer models, they lend themselves very well to computer simulation and approximation. The code for implementation of an approximation scheme comes directly out of the differential equation formats.

Previous versions of this book have been used for a course in mathematical modeling at the University of Wisconsin–Milwaukee for about a dozen years. This is a third-year course requiring a semester of calculus and a semester of matrix theory. The students have come from a number of different disciplines: about one-half from mathematics and others from biology, economics, sociology, and even architecture.

Typically, about half the material in this book was covered in a semester and included very little from Part IV. This was usually reserved for students who presented a project based on the material in one of these later chapters.

Although initially conceived as a textbook, it has evolved into something a bit more extensive. Certain chapters are potentially useful to researchers in ecology, in fisheries sciences, and in biomedicine. Most of the material in the later chapters, although not new, has not appeared in book form previously.

GILBERT G. WALTER
MARTHA CONTRERAS

List of Figures

1

Introduction and Simple Examples

1.1 Mathematical Models

The purpose of a mathematical model is to explain or predict some phenomenon in the "real world." This real world is the one in which measurements and observations are made. These, in turn, may be informal such as the observation that a traffic jam always forms on a certain corner, or they may consist of precise measurements of the outcome of a physics experiment. By themselves, any measurements are meaningless; they must be put into some context to give them sense. This context is a model. Sometimes the model is in form of a verbal or visual model, but often it is, in fact, the mathematical model in question and assumes the form of equations. These equations may then be solved to obtain desired predictions. Of course, there are many cases in which the equations cannot be solved, but are used instead to derive properties of the solutions.

This process is already familiar to most students of calculus and other elementary mathematics courses. The "word problems" in these courses are based on real-world observations (presumably) and a model in the form of equations or integrals is used to make some sort of prediction. In the context of applications of a particular technique, these problems are fairly easy to solve, but become much more difficult if removed from this context, as they are in the real world.

However, the process is often not so straightforward. Rather, the initial formulation of the model may be tentative and must be subsequently refined and corrected to give a more realistic model. Thus, we see that the construction of mathematical models involves a number of steps, as illustrated in Figure 1.1.

The scientist or other real-world observer records his observations and formulates a tentative mathematical model to explain them. This tentative model is then analyzed by solving equations or making other inferences based on it. These inferences are then used to reach conclusions about the

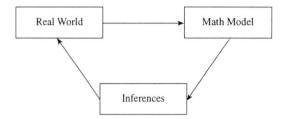

Figure 1.1: Steps in constructing a mathematical model.

real world, which, in turn, are used with further observations to formulate a new model, and the cycle is repeated again.

The most difficult part of this procedure is the first step. Just as with calculus word problems, the construction of models is difficult without a context, i.e., without knowing what sort of model should be used in a particular setting. Fortunately there is a class of models, the *network or flow models*, which are useful in a large number of different real-world settings. They are easier to formulate since they are based on directed graphs which are closely related to the real-world formulation. These directed graphs (digraphs) may themselves constitute the model or they may be refined to weighted digraphs, to Markov chains, or to compartmental models. In this work, we shall concentrate on such models.

We begin with graph theory and its applications and then go on to the other topics. But first we present a few examples to give a feel for what we are going to do.

1.2   Examples of Models

We consider here various examples involving the three levels of network models. We begin with a prototype example which can be modeled as a digraph, as a Markov chain, or as a compartmental model.

Example 1

A beer maker has a small hole in his barrel from which he loses beer at the rate of 10% of the quantity in the barrel each week. He starts off with 64 gallons. How long before it's half gone? Before it's all gone?

The simplest model of this problem is a directed graph in which the two vertices (or nodes or points) I and O correspond to the quantity of beer in the barrel and the quantity outside, respectively.

The arc (i.e., arrow) indicates that there is a flow from inside the barrel to the outside. But beyond observing that the inside amount is decreasing

Figure 1.2a: A digraph model.

and the outside is increasing, we cannot infer much from this model.

If we include the fact that the weekly flow rate from the inside to the outside is 0.1 times $B(t)$, the number of gallons inside the barrel, then the model becomes a *weighted digraph*.

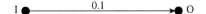

Figure 1.2b: A weighted digraph model.

This weighted digraph may be interpreted in two ways—one involving discrete time and probabilities and the other continuous flows. The first leads to a *Markov chain* and involves following a molecule of beer through the system. The probability that it moves from inside the barrel to outside during a particular week is 0.1. The probability that it stays in the barrel during that week is 0.9. Once outside, it stays there. These probabilities may be represented by a *stochastic digraph* which differs from the weighted digraph of Figure 1.2b in that it includes loops as well.

Figure 1.2c: A stochastic digraph model.

The weights in this stochastic digraph are the *transition probabilities* of the Markov chain models. In these models, time is measured in discrete steps, and the probabilities that the molecule is in one of the two compartments are found at each such time step. Since each molecule must eventually end up outside the barrel, we might ask how many time steps it would take on average. The answer in this case is 10 weeks, but this requires some of the theory which we return to in Chapter 11.

The weight of Figure 1.2b can also be converted into a continuous quantitative model by using the *mass balance equations*. These tell us that the rate of change of level is equal to the difference between the flow rates in and the flow rates out of each compartment. For the inside of the barrel compartment (I), this rate of change (the derivative) becomes

$$dB/dt = -0.1B \qquad (1.1)$$

gallons per week. For the outside, it is

$$dC/dt = 0.1B, \qquad (1.2)$$

where C is the quantity outside (spilled). These two differential equations together with the initial conditions

$$B(0) = 64, \quad C(0) = 0$$

constitute a "compartmental model." Such models are the chief subject of this book.

The next step in modeling, the inference step, involves solving these two equations. This can be done by techniques discussed in Part III. The solution to the first equation is

$$B(t) = 64\,e^{-0.1t}. \qquad (1.3)$$

The equation for spilled beer is then based on this; the solution is

$$C(t) = 64(1 - e^{-0.1t}). \qquad (1.4)$$

We could now check our mathematical model to see how well it compares to the real world. If we had data on the amount of beer in the barrel at several different times, we could compare these to the values predicted by the model. This might lead us to change the value of the parameter 0.1 or perhaps even the functional form of the flow rate.

The solution to the problem originally posed, i.e., to find when the barrel is half-empty, is easy now. We set $B(t)$ to 32 and solve for t ($t = (\ln 2)/0.1 = 6.93$). But the barrel will never empty out completely according to this model! This indicates that a different functional form for the flow rate is more realistic.

Example 2

In this example, a graph-theoretic model is sufficient. However, the model will be a multigraph rather than a digraph.

The city of Milwaukee has three rivers flowing into its harbor. Seven bridges cross the rivers as shown in Figure 1.3. A bridge inspector wants to plan his route to cross each bridge once and return to his office. How can he do so?

A simple model consists in converting the map into a multigraph. This is a structure in which the vertices are joined by one or more undirected

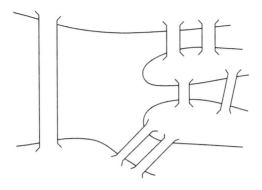

Figure 1.3: A map of part of Milwaukee.

edges. The vertices in the model will correspond to the ends of the bridges, whereas the edges correspond to the bridges themselves. The map is then converted into the model shown in Figure 1.4. From this, it is clear that there is no solution to the problem since any path followed by the inspector must leave each vertex as often as it enters it. Since vertices 1 and 3 each have an odd number of edges adjoining them, this is impossible. Later, we shall see that this is a special case of Euler's theorem.

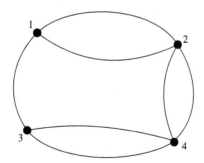

Figure 1.4: The multigraph model of the map.

Example 3

In the game of Russian roulette, a revolver is loaded with a bullet in one of the six chambers, and the others are empty. The player spins the cylinder, holds the revolver to his head, and pulls the trigger. If the gun goes off, he loses; if not, he wins. The probability of winning is 5/6 and the probability of losing is 1/6. How many games could he expect to play before finally losing?

A first model is a stochastic digraph, with weights corresponding to the probabilities of going from one state (W to L) to another in a single play. Thus, a loser stays a loser with probability 1, but a winner has probability 1/6 of becoming a loser and 5/6 of staying a winner. Interpreted this way, the model again becomes a Markov chain, which is similar to that in Example 1. But this example gives a better feel for the notation and terminology.

Figure 1.5: The stochastic digraph for Russian roulette.

We can answer the question about the expected number of games by finding the probability that he plays exactly one game (1/6), exactly two games ((5/6)(1/6)), exactly n games ($(5/6)^{n-1}(1/6)$), since he must win $n-1$ games before losing one. The expected number of games is the average weighted by these probabilities and is given by the infinite series

$$E = 1(1/6) + 2(5/6)(1/6) + \cdots + n(5/6)^{n-1}(1/6) + \cdots.$$

This may be summed by using the fact that

$$1 + 2r + 3r^2 + \cdots + nr^{n-1} + \cdots = (1-r)^{-2}, \tag{1.5}$$

which, in turn, is obtained by differentiation of the geometric series

$$1 + r + r^2 + \cdots + r^n + \cdots = (1-r)^{-1}. \tag{1.6}$$

Hence, $E = (1/6)(1-5/6)^{-2} = 6$.

Example 4

A meadow ecosystem consists of birds, snakes, mice, grass, and insects. The snakes eat the mice and the birds, the insects eat the grass, and both the mice and birds eat the insects and grass. This may be described by the digraph below. It can then be made into a quantitative model by converting it into either a Markov chain or a compartmental model. In the former, we would trace the probabilities that a molecule of, say, carbon passes from one of the compartments to another in a given day. For the latter, we

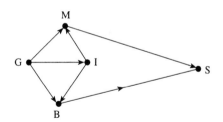

Figure 1.6: The digraph of a meadow ecosystem.

would need the flow rates between the various compartments. These could have various forms and could be linear or nonlinear.

Example 5

A drug is taken orally and becomes absorbed from the stomach into the bloodstream and then is gradually eliminated. The initial dose is 500 mg, and the transfer rate from the stomach to the bloodstream is 10%/hr, and the elimination rate is 5%/hr. How long will it be before 90% of the drug is eliminated? A compartmental model of the flow of the drug is shown in Figure 1.7.

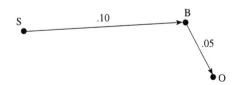

Figure 1.7: A compartmental model of a drug ingestion.

Just as in the first example, this may be converted into a system of differential equations. If we let $X(t)$ and $Y(t)$ be the quantity of the drug in the stomach and bloodstream, respectively, then we obtain the differential equations

$$\frac{dX}{dt} = -0.10X, \qquad X(0) = 500,$$

$$\frac{dY}{dt} = 0.10X - 0.05Y, \quad Y(0) = 0.$$

The solution is

$$X = 500\exp(-0.10t), \quad Y = 1000(\exp(-0.05t) - \exp(-0.10t)).$$

From this, we may deduce that both compartments empty out eventually, but answering the question originally posed, i.e., when 90% of the drug has been eliminated, requires some numerical analysis. For example, the maximum level of the drug in the bloodstream is less than the initial dosage, but how much? We shall return to these questions in Part IV.

At this stage, the interested reader is probably asking: What are they talking about anyway? What exactly are these "graphs," "digraphs," "multigraphs," "compartmental models," "Markov chains," and "systems of differential equations," and what do we do with them? The answers will be the subject of the remaining chapters.

Part I

Structure of Models: Directed Graphs

The use of directed graphs (digraphs) as mathematical models is deceptively simple, involving, as it does, merely drawing a diagram corresponding to the real world problems. The analysis however can become quite difficult and even in some cases, impossible. Nevertheless, their applications are quite widespread and include such areas as scheduling problems, matching problems, food web analysis, the study of hierarchies, map coloring problems, and analysis of games. Their theory is not so widely known as other parts of mathematics and therefore we shall devote some of this part to this theory as well as to applications.

We begin with formal definitions of digraphs (and graphs which we shall treat as special cases of digraphs), and then develop some of their properties. We shall use some of these to study certain applications; however, we shall also use them in the subsequent construction of more complex models.

Many books on graph theory are available, but most rapidly get to questions which are beyond the scope of this book, and which may not be so important in applications. However, the books by O. Ore (1963), G. Chartrand (1977), and F. Roberts (1976) are quite readable and contain some interesting problems which we have omitted.

2

Digraphs and Graphs: Definitions and Examples

In the Introduction, we referred to digraphs or directed graphs several times, but we did not define them. We saw that they are objects that look like those in Figure 2.1.

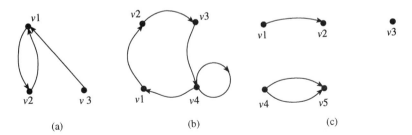

Figure 2.1: Three examples of digraphs.

2.1 Definitions

A digraph consists of a *vertex set* V, a finite, nonempty set, together with a subset A of the Cartesian product $V \times V$ of V with itself, consisting of *arcs*; that is, A consists of ordered pairs of the vertices in V; these correspond to the arrows in the diagram of the digraph. This diagram is what we draw to represent the digraph. In Figure 2.1, diagram (a) has as its vertex set

$$V = \{v_1, v_2, v_3\},$$

and its arc set
$$A = \{(v_1, v_2), (v_2, v_1), (v_3, v_1)\}.$$

We shall refer to the digraph by either the diagram or as the pair $D = (V, A)$.

This definition allows loops as in (b) of the form (v_4, v_4), and also allows the digraphs to be disconnected as in (c). For the most part our digraphs will not contain loops, but they will turn out to be important when we study Markov chains.

A *graph* G is a digraph for which the arc set is symmetric; i.e., it contains (v_j, v_i) whenever it contains (v_i, v_j). The pair of arcs is usually combined into the *edge* $\{v_i, v_j\}$, and hence the graph may be considered to be the pair $G = (V, E)$, where E is the set of all these edges. Thus, a graph may be represented by using its arc set or by its edge set as in Figure 2.2.

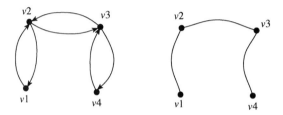

Figure 2.2: Two diagrams for the same graph.

2.2   Examples

Some examples of the applications of digraphs were given in the Introduction. Here are several more examples of both digraphs and graphs.

Example 6

Communications network. A dispatcher for a cab company can communicate one way with each cab, but has two-way communication with a customer. The customer, the dispatcher, and each of the cabs correspond to a vertex, while the line of communication correspond to arcs in the digraph of Figure 2.3.

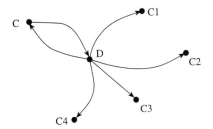

Figure 2.3: The digraph of a communications network.

Example 7

Pecking order. In a certain barnyard, for every pair of chickens either the first is pecked by or it pecks the second chicken. Each of the chickens is a vertex and (c_i, c_j) is an arc if c_i pecks c_j. This leads to a number of natural questions. Is there a unique pecking order? Even if there isn't, is there always a unique leader? A digraph of the pecking order might look like Figure 2.4.

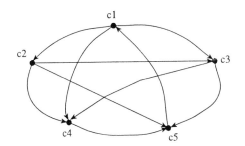

Figure 2.4: A possible pecking order digraph.

Example 8

Scheduling procedure. In a certain factory, there are a number of manufacturing operations that must be done in order to produce a certain product. They are casting, cleaning, grinding, milling, drilling, and inspection. Casting must be done before cleaning, and cleaning must be done before each of the other remaining operations, and inspection must follow all operations. This may be represented by a digraph as shown in Figure 2.5. In order to determine the minimum time needed to produce this product, we consider all paths from the start to the finish. The time needed for the path with longest time is the value for which we are looking. Suppose casting takes 1 hour, cleaning and drilling each 5 minutes, grinding 8 minutes, and milling and inspection each 10 minutes. Then, the operations in the path *ca, cl, gr,*

in take a total of 83 minutes, while those in the path *ca*, *cl*, *mi*, *in* take 85
minutes. There is no way we can schedule the operations to do all of them
in less than 85 minutes. This is a *critical path* which determines the minimum time. There are several possible ways in which the six operations can
be scheduled (ordered), but only one which achieves this minimum time.
It consists of doing the three operations *gr*, *mi*, and *dr* at the same time.
They must, of course, all be preceded by *ca* and *cl*, and followed by *in*.

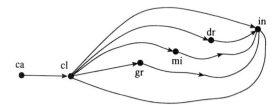

Figure 2.5: An ordinary digraph of operations in a factory.

Example 9

Transportation network. The cities of Min, Mil, Mad, and Chi are joined
by bus lines as shown in Figure 2.6. If the bus drivers go on strike, where
is the best place, if any, to set up a picket line and cut the network to
disrupt the service? The answer, of course, is at any bridge. In graph
theory terms, this means any edge whose removal makes a connected graph
disconnected. Such bridges also arise in questions of orientation of graphs.
In this particular graph, there are no bridges and hence no place to disrupt
the service. There is still always a way to go between any two cities.

Figure 2.6: The graph of a transportation network.

2.3 Problems

1. Express each of the digraphs in Figure 2.1 in the form (V, A), i.e., exhibit the vertex set and the ordered pairs in the arc set.

2. Express the graph in Figure 2.2 in the form (V, E), where E is the edge set.

3. In the communications network, the customer can communicate indirectly with each of the cabs through the dispatcher, but the cabs have no way of calling back. In a more modern system, the customer communicates with an operator, who, in turn, gives orders to a dispatcher with two way communication with each cab. Sketch the corresponding digraph. Which system is better?

4. In the pecking order digraph shown in Example 7, is there a unique pecking order? If not, is there any pecking order of the five chickens which is consistent? (That is, if a pecks b and b pecks c, does a peck c?) If not, give an example of a pecking order of five chickens which is consistent.

5. Digraphs of the type in Example 7 can be used in a number of applications. In paired comparison tests, a person chooses which of a pair of products she or he prefers. For example, among the foods broccoli, kohlrabi, spinach, bok choy, and rutabagas, she or he might prefer broccoli to spinach, bok choy to kohlrabi, etc. Sketch a digraph giving your own preferences for these foods, and answer the same questions as in Problem 4.

6. Digraphs of the type in Example 7 are often called tournaments. Choose a set of teams from the NBA, and construct a hypothetical tournament, in which each pair of teams has a winner. Does this give a unique order?

7. In the scheduling digraph of Example 8, sketch a digraph for each possible scheduling of the six operations. The original digraph gives one scheduling digraph, but there are many other possibilities. For example, the operations may all be scheduled sequentially. Suppose only two of these three operations can be done simultaneously. Find an optimal schedule for this case.

8. In the transportation network problem of Example 9, there is no best solution without other considerations such as the number of people served, the distances, etc. For a general transportation network, try to formulate conditions under which cutting one link would disrupt the network. (The corresponding graph theoretic concept is the existence of a "bridge" whose removal makes the graph disconnected. Such bridges also arise in questions of orientation of graphs considered in Chapter 4.)

3

A Little Simple Graph Theory

In this chapter, we concentrate on some of the theoretical concepts and results in graph theory. These are usually intuitive and easy to understand, particularly in terms of the associated diagram. However, there are a lot of definitions to keep straight; furthermore there is no unanimity in the literature about these definitions, so other references may not be helpful. We present the proofs of theorems in some cases as well. These are more difficult than the other materials and can be omitted on first reading.

3.1 Isomorphic Graphs and Digraphs

Two graphs or digraphs with the same vertex set are considered the same (isomorphic) if when the vertices are permuted, the edges or arcs are too. More precisely, if $G_1 = (V_1, E_1)$ and if $G_2 = (V_2, E_2)$, then G_1 and G_2 are isomorphic if there is a bijection

$$\phi : V_1 \rightarrow V_2$$

such that $\{u, v\}$ is an edge in E_1 if and only if $\{\phi(u), \phi(v)\}$ is an edge in E_2. The bijection will usually be a permutation of the vertices in V_1, but this definition allows us to change the names as well.

The two graphs shown in Figure 3.1 are isomorphic, but those in Figure 3.2 are not.

Some basic results are as follows.

PROPOSITION 3.1

(a) *Isomorphic graphs have the same number of edges and vertices.*

(b) *The number of edges leaving each vertex is invariant under isomorphism.*

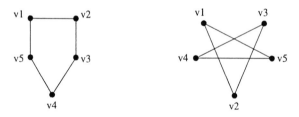

Figure 3.1: Two isomorphic graphs.

Figure 3.2: Two nonisomorphic graphs.

(c)   Isomorphism is an equivalence relation for graphs.

The proofs are obtained by writing down the statements precisely, and are left to the interested reader. An equivalence relation requires that (i) $G_3 \simeq G_3$, (ii) if $G_1 \simeq G_2$, then $G_2 \simeq G_1$, (iii) if $G_1 \simeq G_2$ and $G_2 \simeq G_3$, then $G_1 \simeq G_3$ (reflexive, symmetric and transitive).

We will not distinguish between isomorphic graphs.

Digraphs also have an associated concept of isomorphism, except that arcs instead of edges have to be used. Proposition 3.1 holds in this case as well if each occurrence of the word edges is replaced by arcs.

It is now a simple matter (theoretically) to list all the graphs of a given order (i.e., of a given number of vertices) if we count isomorphic graphs only once. For example, there are clearly four graphs of order 3, but finding the number of graphs of order 4 is a little more difficult. In the case of digraphs, things are more complicated. If a digraph has three vertices and two arcs, it may not be isomorphic to another with the same number of arcs and vertices. For graphs however, there is only one with three vertices and two edges.

Problems 3.1

1. (For the interested reader.) Prove Proposition 3.1. (Hint: for (b) show that $d(u)$, the number of edges adjoining the vertex u, satisfies

$d(u) \leq d(\phi(u))$, where ϕ is the isomorphism. Since ϕ is a bijection, the opposite inequality also holds.)

2. State and illustrate a version of Proposition 3.1 suitable for digraphs.

3. Construct all possible graphs of order 3 and of order 4.

4. Construct all possible digraphs of order 3, without loops.

5. Show that the graphs in Figure 3.2 are not isomorphic.

6. The following graphs have the same number of vertices and the same number of edges adjoining each vertex. Show they are not isomorphic. (You have to go back to the definitions since the proposition doesn't apply. Use the fact that adjacent vertices, i.e., those joined by an edge, must be mapped onto adjacent vertices by ϕ.)

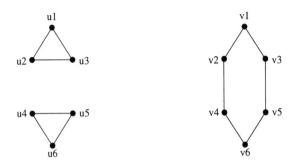

Figure 3.3: Two nonisomorphic graphs.

7. Construct six nonisomorphic digraphs of order 4 with four arcs.

3.2 Connected Graphs and Digraphs

Intuitively, a graph is connected if it is all one piece, i.e., if we can draw a path from any vertex to any other following the edges. The same is true for digraphs, but we have to be sure we are going in the directions of the arcs.

DEFINITION 3.1   A path in a digraph $D = (V, A)$ (or in a graph $G = (V, E)$) is a sequence of alternating vertices and arcs in A (or edges in E)

$$u_1, a_1, u_2, a_2, u_3, \ldots, a_n, u_n$$

such that $a_i = (u_i, u_{i+1})$ (or $\{u_i, u_{i+1}\}$). For brevity, we frequently list only the vertices in describing the path.

Caution: The path in a graph is not the same as the path in the corresponding digraph.

Example 3.1

In the first graph of Figure 3.3, $u_1, \{u_1, u_2\}, u_2, \{u_2, u_4\}, u_4$ is not a path, but the same sequence with u_4 replaced by u_3 is a path.

DEFINITION 3.2   A path in a digraph is simple *if it does not repeat any vertex, is* closed *if $u_{n+1} = u_1$, and is a* cycle *if it is both closed and simple (except for the first and last vertex). It is* Eulerian *if it uses each arc exactly once, and is* complete *if it uses each vertex in V. The* length *of a path is the number of arcs in it.*

The same definitions hold for paths in graphs except that arc must be replaced by edge. A closed and simple path will be called a cycle whether composed of edges or arcs. This will cause a little confusion later.

In the graph of Figure 3.4, the path a, b, c, e, h is simple, whereas the path b, c, d, e, c is not. The path c, d, e, c is closed and is a cycle; but b, c, d, e, c, b is also closed but is not a cycle. There are no Eulerian paths, but there are complete paths such as $a, b, f, g, f, e, c, d, e, h$.

Figure 3.4: A graph for illustrating definitions.

We now are able to introduce the formal definition of connectedness of a graph and digraph.

DEFINITION 3.3   A graph G is connected *if for each pair of vertices u, v in V, there is a path from u to v. A digraph D is* strongly connected *if for each u, v in V, there is a path from u to v and one from v to u. It is* weakly connected *if the associated graph G, obtained by replacing each arc (u, v) by an edge $\{u, v\}$, is connected.*

Clearly, a strongly connected digraph is also weakly connected. The graph in Figure 3.4 is connected, whereas the first graph in Figure 3.3 is not connected. In Figure 3.5, we show examples of digraphs with various connectedness properties.

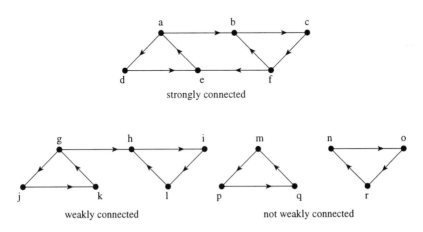

Figure 3.5: Connectivity of digraphs.

PROPOSITION 3.2
If there is a path from u to v, then there is a simple path.

PROOF Among all paths from u to v, there is one of shortest length, $u = u_1, u_2, \ldots, u_{n+1} = v$. If this is not simple, then there is at least one repeated vertex, say $u_i = u_{i+k}$, in the path. Then, $u_1, \ldots, u_i, u_{i+k+1}, \ldots, u_{n+1}$ is a path from u to v of shorter length. Thus, this shortest path must be simple. ∎

PROPOSITION 3.3
A graph G is connected if and only if it contains a complete path. A digraph D is strongly connected if and only if it contains a complete closed path.

PROOF We prove only the part about strong connectivity; the other is very similar. Let D be strongly connected with vertices $u_1, u_2, \ldots, u_n$. There is a path from u_1 to u_2, from u_2 to $u_3, \ldots$, and from u_n to u_1. Joined together, they form a complete path, which, since it begins and ends with u_1, is closed.

Let $u_1, u_2, \ldots, u_n, u_1$ be a complete closed path in D. Each pair of vertices appears in some position on this path, say at u_i and u_k. If $i < k$, then $u_i, \ldots, u_k$ is a subpath of this path and $u_k, \ldots, u_1, u_2, \ldots, u_i$ is a path since the original path was closed. Hence, there is a path from u_i to u_k and one from u_k to u_i and, thus, D is strongly connected. ∎

DEFINITION 3.4   Let $G = (V, E)$; then, a subset H of G is a subgraph of G if $H = (V_H, E_H)$ is a graph and $V_H \subseteq V$ and $E_H \subseteq E$.

The same definition holds for digraphs if the edge sets are replaced by the arc sets. We shall use the convention that the term subgraph can refer either to a graph or a digraph.

DEFINITION 3.5   A component of a graph G is a maximal connected subgraph; a strong component of a digraph D is a maximal strongly connected subgraph of D.

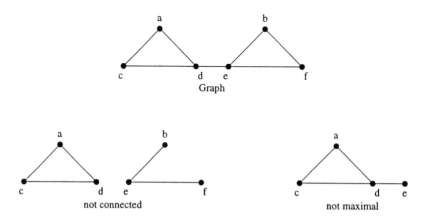

Figure 3.6: Subgraphs which are not components.

PROPOSITION 3.4

The set of all its connected components partition a graph G; the set of all its strongly connected components partition a digraph D.

PROOF   The proof is given only for graphs; that for digraphs is similar.

Let u_1 be any vertex in V; then, $\{u_1\}$ as a subgraph is connected. Add a vertex u_2 if possible such that $\{u_1, u_2\}$ is an edge; then, add another u_3 such that $\{u_i, u_3\}$, $i = 1$ or 2, is an edge and keep adding vertices until it is no longer possible. The resulting set of vertices with their associated edges is connected and is maximal. Hence, each vertex belongs to a component.

Let H and K be two components and let u be in the intersection of their vertex sets; let u_1 be in H and u_2 be in K. Then, there is a path from u_1 to u and from u to u_2. Hence, $H \cup K$ is connected; this contradicts maximality and, therefore, u can belong to only one of K and H.   ∎

DEFINITION 3.6   *The* condensation D^* *of a digraph* $D = (V, A)$ *is a digraph* $D^* = (V^*, A^*)$, *where* V^* *consists of the strongly connected components of* D *and* A^* *is obtained from arcs between components, i.e.,* (K_i, K_j) *is an arc in* A^* *if and only if there is a vertex* u *in* K_i *and* v *in* K_j *such that* (u, v) *is in* A.

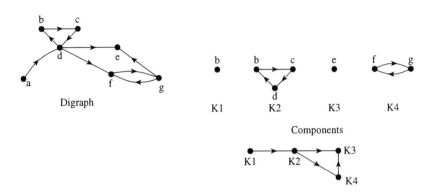

Figure 3.7: A digraph and its condensation.

In the example of Figure 3.7, the condensation is acyclic. This property is useful for the construction of vertex bases, which we shall take up in the next section. Other uses of these components will appear in Part II and Part III.

Problems 3.2

1. Let G be a connected graph of order n; what is the maximum and minimum number of edges in G?

2. Let D be a strongly connected digraph of order n; what is the minimum number of arcs in D?

3. Find the strong components and use them to construct the condensation of the digraph in Figure 3.8.

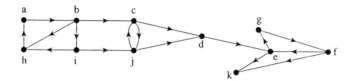

Figure 3.8: Digraph for problem 3.

4. Find an example of a strongly connected digraph which does not have a complete cycle? Why doesn't this contradict Proposition 3.3?

5. Find a simple path from vertex a to each of the other vertices in the digraph of Figure 3.7. Do the same for the digraph of Figure 3.8.

6. Are there any other vertices in the digraph of Figure 3.8 that have simple paths to all vertices? Which ones?

7. Would you expect a pecking order digraph to be strongly connected? Why (not)?

8. Suppose a transportation network is composed of one-way roads. Why must the corresponding digraph be strongly connected?

4

Orientation of Graphs and Related Properties

A graph can be converted into a digraph by assigning a direction to each of its edges, i.e., by giving it an orientation. In this chapter, we discuss desirable properties of such orientations. These include the Eulerian orientation of multigraphs and strongly connected orientations of graphs. Related concepts are those of vertex basis, i.e., a minimal set of vertices from which it is possible to reach all vertices, and spanning trees. Various ways of constructing the latter are introduced; in particular, the greedy algorithm, which leads to spanning tree which is optimal in some cases. If the original graph is a weighted graph, this is optimal in the sense of minimum cost.

4.1 Vertex Basis

In a digraph corresponding to a communication network, one is frequently interested in finding a set of vertices from which one can reach every other vertex. Moreover, one would like this set to be minimized in order to spread a message as efficiently as possible. The same is true in the digraph of a military hierarchy. The generals want to be able to transmit their orders down the ranks to the privates efficiently. In a computer file system, the directory tree is a digraph. The root directory is the one from which all the others can be reached. In each of these cases, we are first interested in finding a vertex basis of the digraph.

DEFINITION 4.1   *A vertex basis of a digraph D is a set $B \subset V$ such that for each u in V, there is a path to u from some v in B, and the set B is minimal.*

THEOREM 4.1

If D is acyclic (i.e., contains no cycles), then it has a vertex basis B consisting of all vertices with no incoming arcs.

PROOF   Let B be such a set. Then, B is a subset of each vertex basis. If $v \notin B$, then v has an incoming arc, say (u_1, v). If $u_1 \in B$, then there is a path from B to v; if $u_1 \notin B$, there is an arc (u_2, u_1). We repeat this argument to obtain a path $u_k, u_{k-1}, \ldots, u_1, v$, all of those elements are distinct. If they were not, D would contain a closed path and, hence, a cycle. Eventually, V will be exhausted and the last vertex added would have no incoming arcs and would, therefore, belong to B. Clearly, B is also minimal and, hence, is a vertex basis.   ∎

Digraphs which are not acyclic also have vertex bases, but they may not be unique. They are found by considering the condensation D^* of the digraph D.

THEOREM 4.2

The condensation D^* is always acyclic.

PROOF   Suppose D^* has a cycle $K_1, \ldots, K_n, K_1$ of components of D. Let $u_i \in K_i$, $u_j \in K_j$, $i < j$, for two components on this cycle. Then, there is a path u_1 to u_j in D and from u_j to u_i, and $K_i \cup K_j$ must be strongly connected. By the maximality of both, $K_i = K_i \cup K_j = K_j$, a contradiction. Hence, D^* had no cycles.   ∎

COROLLARY 4.3

Let D be a digraph with condensation D^*; let $B^* = \{K_1, K_2, \ldots, K_d\}$ be the vertex basis of D^*, and let $B = \{u_1, u_2, \ldots, u_d\}$, where $u_i \in K_i$; then, B is a vertex basis of D.

PROOF   Let $K_i \in B^*$; then, there is a path in D from $u_i \in K_i$ to all other vertices in K_i. There is also a path in D^* from some $K_i \in B^*$ to any other component K. This gives a path in D from u_i to any $u \in K$.   ∎

Problems 4.1

1. Find a vertex basis of the digraphs in Figure 4.1.

2. Is the converse to Theorem 4.1 true? (i.e., is D acyclic if B is a vertex basis?)

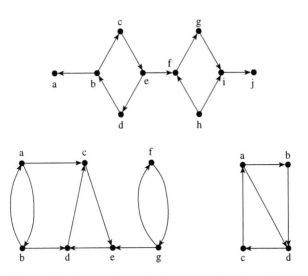

Figure 4.1: Some digraphs for vertex basis problem.

3. Explain how knowledge of the vertex basis might be useful in a model of a communications network, in a model of a computer file systems, and in a model of pollution runoff in a watershed of a river.

4. In an ecosystem, one model that is usually constructed is a digraph which follows the flow of energy through the various components of the system. The vertex basis consists of the primary producers, which convert the sun's energy into a form usable by other components. What might the consequence be of destroying some of the vertex basis?

4.2 Multigraphs

If vertices are joined by more than an edge, we have a **multigraph**. The definition of edge has to be changed slightly to include an index as well: $E = \{(\{v_i, v_j\}, n)\}$ where $n \in I$ some index set. Thus, the first multigraph in Figure 4.2 would have $E = \{((\{a, b\}), 1), (\{a, b\}, 2), (\{b, c\}, 1), (\{a, c\}, 1), (\{a, d\}, 1), (\{a, d\}, 2)\}$.

The definitions associated with paths in graphs carry over to multigraphs. In particular, a path is Eulerian if it includes each edge exactly once. The multigraph itself is Eulerian if it has a closed Eulerian path. It is traversable if it has a Eulerian path which is not necessarily closed.

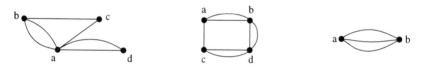

Figure 4.2: Some examples of multigraphs.

The problem about the Milwaukee River bridges in Chapter 1 (example 2) may be converted into a multigraph in which each of the bridges is an edge (Figure 1.4). The bridge inspector's tour must be a closed Eulerian path. In this case, an experiment can lead you to the conclusion that there is no such path and to the proof of the following theorem. It uses the concept of the degree of a vertex, $d(u)$, which is the number of edges adjoining u.

THEOREM 4.4
A multigraph M is Eulerian if and only if M is connected and each vertex has even degree.

PROOF Suppose M has a (complete) closed Eulerian path v_1, e_1, v_2, $e_2, \ldots, v_n, e_n, v_1$. Then, each v_i in path has edges e_{i-1} coming in and e_i leaving ($e_n = e_o$). Since each vertex occurs in the path a finite number of times, $d(v_i)$ is even.

Suppose now that each $d(v_i)$ is even and M is connected. Start a path from any $u_1 \in V$ and add alternate edges and vertices as long as possible without repeated edges; $P = \{u_1, e_1, u_2, e_2, \ldots, u_n, e_n, v\}$. The last vertex v must have appeared previously in the path since otherwise, $d(v) = 1$; in fact, $v = u_j$ is the only possible choice which makes $d(u_1)$ even. Hence, P is a closed path. If it is not Eulerian, then there is an edge not in P but with one of its vertices in P, say u_j. Otherwise, M would not be connected.

Now we form a submultigraph with the same vertices as M but including only unused edges. Let H_1 be the component containing u_j and repeat the argument to get a closed path P_1 in H_1 which contains an edge adjacent to u_j. Combine the two paths to get a longer one. Repeat until all edges are used. ∎

Figure 4.3 is an example illustrating to illustrate the proof (with a graph).

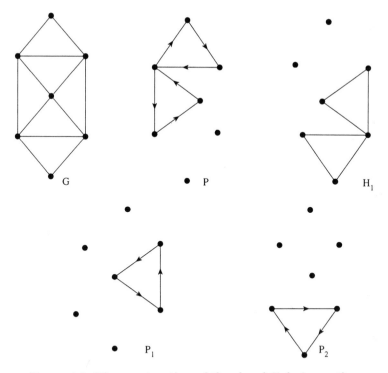

Figure 4.3: The construction of the closed Eulerian path.

COROLLARY 4.5
A multigraph M is traversable if M is connected and has exactly two vertices with odd degree.

PROOF   Let u and v be the two vertices with odd degree and add an edge $\{u, v\}$. The new multigraph has only vertices with even degree and, hence, has a closed Eulerian path.   ∎

In the case of a graph, the directions obtained from the closed Eulerian path determine a digraph, which must be strongly connected. That is, our construction gives us a strongly connected orientation.

Problems 4.2

1. Show that the multigraph of the Milwaukee River bridges is not Eulerian but is traversable.

2. The original multigraph problem considered by Euler involved the bridges of Koenigsberg. They led to the multigraphs in Figure 4.4.

Show that it is not traversable.

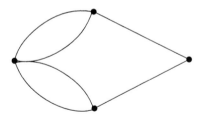

Figure 4.4: The multigraph for the bridges of Koenigsberg problem.

3. A house has the floor plan shown in Figure 4.5, where the gaps are
doors to rooms. Is it possible to follow a path which goes through
each door exactly once?

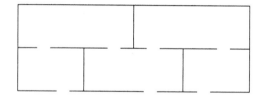

Figure 4.5: A floor plan leading to a multigraph.

4. Which of the graphs in Figure 4.6 are Eulerian? Which traversable?
Find a complete Eulerian path if possible.

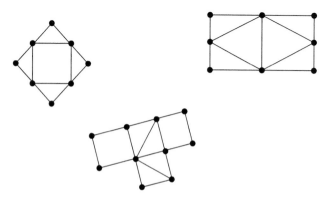

Figure 4.6: Some graphs for checking for Eulerian.

5. Do non-Eulerian graphs ever have a strongly connected orientation? (Consider problem 1.)

6. Do connected graphs always have a strongly connected orientation? If not, give a counterexample.

4.3   Orientation of Graphs

We return to the orientation problem which we did not resolve for non-Eulerian graphs. There are some graphs that clearly do not have a strongly connected orientation, e.g., those that contain a **bridge** as in Figure 4.7. Such graphs arise as maps of cities with a single bridge over a river. The orientation is a one-way street assignment. If we want to get from any part of the city to any other, we need a strongly connected orientation. This is impossible with a bridge, since, whatever the orientation of the bridge, once we cross it, we cannot get back.

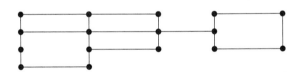

Figure 4.7: A graph with a bridge.

DEFINITION 4.2   *An edge e of a connected graph G, is a bridge if $G - \{e\}$ is not connected.*

PROPOSITION 4.6
An edge e is a bridge if and only if e is not on any cycle.

PROOF   Suppose e is part of cycle $u_1, \ldots, u_i, e, u_{i+1}, \ldots, u_1$. Since G is connected for each $u, v \in V$, there is a path from u to v. If e is not in this path for all u and v, then $G - \{e\}$ is connected. If e is in this path, then we go around the cycle in the other direction when we hit e to get another path from u to v. Hence, $G - \{e\}$ is connected in this case too and e is not a bridge. If $e = \{u, v\}$ is not a bridge, then $G - \{e\}$ has a path and, hence, a simple path from v to u and, therefore, $u, v, \ldots, u$ is a cycle. ∎

THEOREM 4.7
A connected graph $G = (V, E)$ is orientable if and only if it has no bridges.

PROOF   If G has a bridge, it is clearly not orientable. Suppose then that G has no bridges and, hence, by Proposition 4.6, each edge lies on a cycle C. We orient the edges in C in the directions given by the cycle. If there is $u \notin C$, then there is an edge from some u to some $v \in C$, and, hence, another cycle C_1 containing $\{u, v\}$. We orient C_1 in a direction consistent with C. This procedure is continued until all vertices in V are on one of these cycles. If at some stage the cycle C_k contains an edge e on some previous cycle C_j with an incorrect orientation, then C_k is modified to include all edges in C_j except e. These now have the correct orientation. The graph consisting of V with the edges in any one of the cycles has a strongly connected orientation. The remaining edges can be oriented in any way, since this cannot destroy the strong connectivity.   ∎

Problems 4.3

1. Check whether the graphs in Figure 4.8 have bridges.

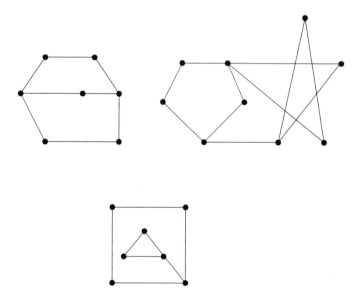

Figure 4.8: Graphs which may have bridges.

2. Find a cycle containing each edge for those graphs which have no bridge.

3. Find a strongly connected orientation by imitating the proof of Theorem 4.7.

4.4   Spanning Trees

We saw in Section 4.3 how to construct an orientation for certain graphs without bridges. In this section, we look for subgraphs composed entirely of bridges. These are important in applications where we need to avoid redundancy. In particular, they can be used to solve the **minimum connector** problem, i.e., that of finding a complete subgraph whose cost is minimum.

DEFINITION 4.3   A graph T is a tree if it is connected and has no cycles.

Some examples of trees are given in Figure 4.9.

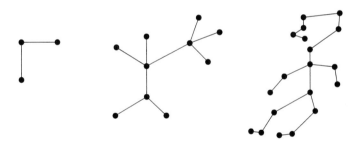

Figure 4.9: Some trees.

PROPOSITION 4.8
For each pair of vertices in a tree, there is a unique path joining them.

This is clear since if there were two distinct paths, there would be a cycle.

PROPOSITION 4.9
A tree of order p has p − 1 edges.

The proof is by induction on *p*. There is at least one vertex of degree 1. Remove it and the adjoining edge for the induction step.

DEFINITION 4.4   A subgraph T of connected graph G is a spanning tree if it is complete with the same vertex set and is a tree.

By comparing the spanning trees of a graph, we can solve the minimum connection problem. There are several methods for finding such spanning trees. One is **breadth first search** which we illustrate with the graph in Figure 4.10. Its resulting spanning tree is shown in Figure 4.11.

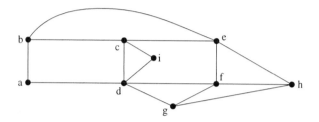

Figure 4.10: A graph to illustrate spanning tree construction.

We start with any vertex, say b, and add all edges and vertices adjacent to b: $\{b, a\}$, $\{b, c\}$, and $\{b, e\}$. Now, do the same for each of the adjacent vertices a, c, and e, adding all edges not covered before while avoiding cycles. This gives new edges $\{a, d\}$, $\{c, i\}$, $\{e, f\}$, and $\{e, h\}$. Repeat with each of the new vertices d, i, f, h and continue until all vertices are exhausted.

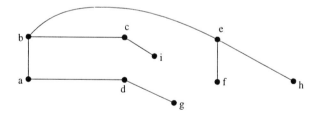

Figure 4.11: The breadth first spanning tree for the graph of Figure 4.10.

Another common method is the **depth first search** spanning tree. Again, we illustrate the procedure with the same graph. We start at b and add any edge and vertex adjacent to b, say $\{b, a\}$; then, go to a and add an edge, say $\{a, d\}$, then to d and add edge $\{d, c\}$, to c and add edge $\{c, i\}$. But then we're stuck, since we must avoid cycles, so we back up to the previous (i.e., deepest) vertex, from which we can go successively to e, h, f, and g.

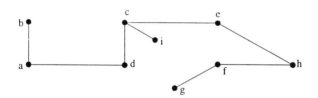

Figure 4.12: The depth first search spanning tree for the graph of Figure 4.10.

This depth first search is another procedure that can be used in constructing a strongly connected orientation for a connected graph with no bridges. The edges in the tree are oriented in the direction of construction (starting from b in the example), i.e., consistent with an ordering of the vertices given by the order in which they are added, b, a, d, c, i, e, h, f, g. The remaining edges are oriented in the direction from the later vertices to the earlier ones. In the case of Figure 4.10, the orientation is given in Figure 4.13.

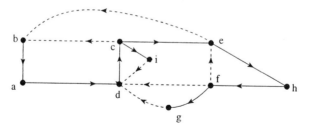

Figure 4.13: Orientation arising from depth first search spanning tree. (Dotted arcs are not part of spanning tree.)

A tree, spanning or otherwise, always has a unique simple path between every pair of vertices. If it is given an orientation with a vertex basis consisting of a single vertex, then the digraph is called a *rooted tree* and the vertex is this root. In Figure 4.12 with the orientation described above, vertex b will be the root.

Such rooted trees are often used to describe organizational structures, geneological charts, file systems, and decision trees. The various directories and subdirectories of a computer constitute a rooted tree.

Example 4.1

Suppose we have six coins, each of which is counterfeit and lighter than the others but which looks the same. We label the coins 1, 2, 3, 4, 5, and

6 and compare $\{1, 2, 3\}$ to $\{4, 5, 6\}$ by using a balance scale. If $\{1, 2, 3\}$ is lighter, then compare $\{1\}$ and $\{3\}$. If either one is lighter, that's it. If both are the same, it's $\{2\}$. This is described by the rooted tree in Figure 4.14.

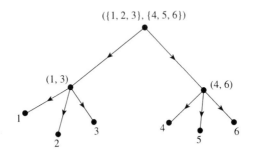

Figure 4.14: The rooted tree of Example 4.1. The symbol (A, B) means A is compared to B. The arc to the left is followed if A is lighter, the other when B is. Notice that this requires only two weighings instead of five.

Problems 4.4

1. Find a breadth first spanning tree for the graphs in Figure 4.15.

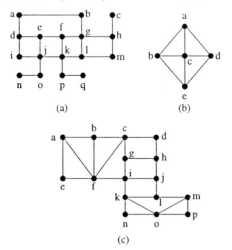

Figure 4.15: Graphs for spanning tree problem.

2. Use the depth first search to find a spanning tree for the three graphs in problem 1.

3. Use problem 2 to find a strongly connected orientation for all applicable graphs.

4. Find a spanning tree of the graph in Figure 4.15b that is different than the ones you found in problems 1 and 2.

5. Do two spanning trees of a graph always have a common edge? Prove or give a counterexample.

6. Show how to construct a rooted tree beginning with any vertex in a tree. Is it unique? Apply your technique to the graphs in Figure 4.16.

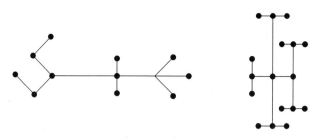

Figure 4.16: Two trees for building rooted trees.

7. Use a decision tree to compare 12 coins, 1 of which is counterfeit and lighter, in no more than 3 weighings on a balance scale.

8. Prove Proposition 4.8.

4.5 Minimum Connector Problem

There are many possible ways of constructing spanning trees for a given connected graph. However, we often need to find an "optimal" spanning tree, one with minimal cost. For example, in a railroad network joining a number of cities, we should like to keep those portions which connect all the cities but are least expensive to maintain. This requires that we prescribe a cost or weight associated with each edge.

DEFINITION 4.5   A weighted graph $G = (V, E, w)$ is a graph (V, E) with an associated weight function $w : E \rightarrow R^+$; i.e., that associates with each edge a positive weight.

In a communications network, $w(e)$ might be the cost in maintaining e; in a road network, the length of e; in a competition graph, the strength of the competition.

Weighted graphs are usually represented by diagrams with the weights next to the edges (see Figure 4.17.

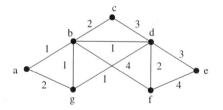

Figure 4.17: A weighted graph.

Similarly, a weighted digraph is defined as one whose arcs each has an associated weight.

The minimum connector problem involves finding a spanning tree of a weighted graph with minimal total weight. For example, the two spanning trees shown in Figure 4.18 have total weights $2+1+2+3+3+4 = 15$ and $2+1+1+2+2+3 = 11$ so that Figure 4.18b is preferred. But is it the best? This seems to be a formidable problem if the graph is at all large. Fortunately, there are a number of simple algorithms that enable us to find the optimum spanning tree, the simplest of which is the *greedy algorithm*. This algorithm says to choose the least expensive choice at each stage.

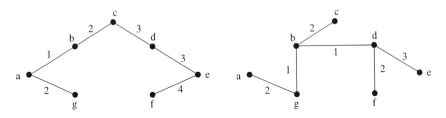

Figure 4.18: Two spanning trees.

Let $G = (V, E, w)$ be a weighted connected graph of order p. The $p-1$ edges of the spanning tree are chosen as follows:

e_1 is an edge with minimum weight in G,

e_2 is an edge with minimum weight in $G - \{e_1\}$,

e_3 is an edge with minimum weight in $G - \{e_1, e_2\}$ such that e_1, e_2, and e_3 are not a cycle,

e_{p-1} is an edge with minimum weight in $G - \{e_1, e_2, \ldots, e_{p-2}\}$ such that $\{e_1, e_2, \ldots, e_{p-1}\}$ contains no cycle.

In the example of Figure 4.17, we may take $e_1 = \{a, b\}$, $e_2 = \{b, g\}$, and $e_3 = \{b, d\}$, all of which have weight 1. We cannot include $\{d, g\}$ because that would give us a cycle. We take $e_4 = \{b, c\}$, $e_5 = \{d, f\}$, and $e_6 = \{d, e\}$ to give us the required $7 - 1 = 6$ edges with total weight 10, which clearly has minimum total weight and is a spanning tree in this case.

THEOREM 4.10

Let $G = (V, E, w)$ be a weighted connected graph; then, the greedy algorithm gives a minimum weight spanning tree.

PROOF Let T be the graph with the $p-1$ edges described in the algorithm and let T_0 be a minimum weight spanning tree. If $T_0 \neq T$, there is a first $e_i \in T$ such that $e_1 \notin T_0$. Adjoin e_i to T_0 to obtain a graph G_0 containing p edges. Since all trees have $p - 1$ edges, G_0 contains a cycle. Therefore there is an edge $e_0 \in T_0$ and on this cycle, but $e_0 \notin T$. We replace e_0 by e_i to get a new spanning tree $T_1 = G_0 - e_0$ whose total weight is $w(T_1) = w(T_0) + w(e_i) - w(e_0) \geq w(T_0)$ and, hence, $w(e_i) \geq w(e_0)$. On the other hand, $w(e_i) \leq w(e)$ for all $e \neq e_1, e_2, \ldots, e_{i-1}$ such that $e, e_1, \ldots, e_{i-1}$ do not contain a cycle. In particular, e_0 is one such edge since together with $e - 1, \ldots, e_{i-1}$, it is part of a tree. Hence, $w(e_i) \leq w(e_0)$ and therefore $w(T_0) = w(T_1)$, so that we have succeeded in replacing one edge in T_0 by one edge in T without changing the total weight. We now repeat this with T_1 and the next edge in T until all the edges have been shifted. Then $w(T) = w(T_0)$.

If T is not a spanning graph, its $p - 1$ edges would adjoin at most $p - 1$ vertices and therefore would have to contain a cycle. If T is not connected, its $p - 1$ edges would have to be distributed among the components of T, one of which would have to contain a cycle. Hence, T is a spanning tree. ∎

Problems 4.5

1. Find a different minimal weight spanning tree for the graph of Figure 4.17.

2. Find a minimal weight spanning tree for each of the weighted graphs in Figure 4.19.

3. A maximal weight spanning tree is found the same way but uses maximum weight. Find such a tree for the weighted graphs in Problem 2.

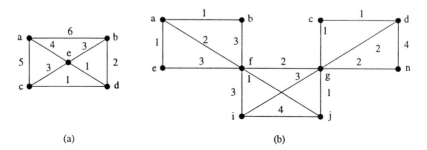

(a)   (b)

Figure 4.19: Two weighted graphs for spanning tree construction.

4. A related problem is to find the shortest path from a fixed vertex to
   all the others in a weighted graph. Show that the shortest path from
   vertex a to the other vertices is not along the minimal spanning tree
   in the weighted graph of Figure 4.19a.

5. Another related problem is the *traveling salesman problem* in which
   the shortest complete closed path is found. Find the shortest (in the
   sense of minimum weight) closed path that includes all the vertices
   in the same weighted graph of Figure 4.19a.

5

Tournaments

The directed graphs in which every pair of vertices has exactly one arc joining them is called a tournament. They are used, e.g., in experiments involving paired comparisons, in round robin tournaments in which each player plays every other one, in studying pecking order in a barnyard or in an organization. Some natural questions that arise with these digraphs are: (i) Is there always a winner? (ii) Is there an ordering of the players determined by the tournament? (iii) If so, is it unique?

5.1 Definitions and Basic Results

We first repeat the definition is a more formal way and then give some simple properties.

DEFINITION 5.1 A tournament *is a digraph $T = (V, A)$ such that for each pair $u, v \in V$, there is either an arc $(u, v) \in A$ or $(v, u) \in A$ but not both.*

Tournaments can become quite complicated if they contain more than a few vertices, but in the case of order 3 and 4, we can list all possibilities (see Figure 5.1). The fact that they are all distinct (i.e., not isomorphic) can be shown by appealing to the digraph version of Proposition 3.1b. The number of arcs leaving the various vertices is different in all the digraphs listed in the figure. For example, in Figure 5.1a, there is one arc leaving each vertex, whereas in Figure 5.1b, there is a vertex with two arcs leaving it, one with one arc, and one with none, so they can't be isomorphic.

One simple property of a tournament is the construction of a larger tournament merely by adding a vertex and arcs (in either direction) to all the other vertices. The size of a tournament can also be reduced by

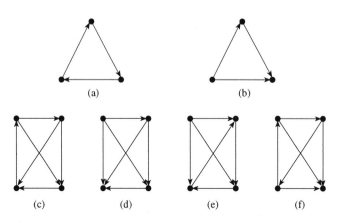

Figure 5.1: All tournaments with three or four vertices.

removing any vertex and its adjacent arcs. It is also easy to see that a tournament with n players (vertices) has $n(n-1)/2$ games (arcs).

DEFINITION 5.2   The score $s(v)$ of a vertex v in a tournament is the number of arcs leaving v (the number of victories of player v). The score sequence of a tournament is the sequence of scores of all vertices in nonincreasing order.

The score sequences of the examples in Figure 5.1 are (a) 1, 1, 1, (b) 2, 1, 0, (c) 2, 2, 1, 1, (d) 3, 2, 1, 0, (e) 3, 1, 1, 1, (f) 2, 2, 2, 0. Now, we are able to indicate in what sense the player with the maximum score is a winner.

PROPOSITION 5.1
Let v be a vertex with a maximum score in the tournament $T = (V, A)$; then, for each $u \in V$, either v beats u or it beats some w that beats u.

PROOF   If v does not beat u, then $(u, v) \in A$. If v beats $v_1, v_2, \ldots, v_m$, then u could not beat all of these since then its score would be greater than that of v. So one of these v_i must beat u. This is the w we need.   ∎

It is easy to check that these conditions hold in the tournaments in Figure 5.1. That it should hold in the tournament in Figure 5.2 is more surprising, since the score sequence is 2, 2, 2, 2, 2. But, for example, a beats b and

c, and b beats d, while c beats e; thus, it satisfies the conclusion for the vertex a. The other vertices can be checked in the same way.

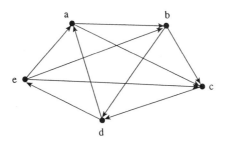

Figure 5.2: A tournament with five players.

We now consider the ranking of our players. Is there always a natural ranking in which u_1 beats u_2, u_2 beats u_3, u_3 beats u_4, ..., etc.? An answer is given by the following proposition.

PROPOSITION 5.2
Each tournament T has a complete simple path.

PROOF We prove it by induction on the number of vertices in T. It is clearly true if T has two or three vertices. Suppose T has $(n+1)$ vertices; then remove one of the vertices, say v, and its adjacent arcs. The remaining vertices then constitute a tournament and by the induction hypothesis must have a complete simple path $v_1, v_2, \ldots, v_n$. Let v_i be the first vertex such that $(v, v_i) \in A$. If $v_i = v_1$, we have our path. If $v_i \neq v_1$, then $(v, v_{i-1}) \in A$ and the path $v_1, v_2, \ldots, v_{i-1}, v, v_{i+1}, \ldots, v_n$ is simple and complete. If there are no vertices such that $(v, v_i) \in A$ then $(v_n, v) \in A$ and again we have our path. (See Figure 5.3.) ∎

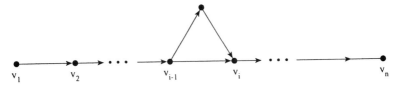

Figure 5.3: The path constructed in Proposition 5.2.

In the example in Figure 5.2 a, b, c, d, e is a complete simple path, but then so is b, d, e, a, c and many others. Therefore, it appears that the ordering given by such a path is generally not too useful, and we might ask under what conditions we get a unique complete simple path? In Figure 5.1, tournament (d) has a unique complete simple path, whereas tournament (e) has several.

5.2   Transitive Tournaments

In order to obtain uniqueness in the ordering, we must add another condition, that of transitivity, which we shall see is exactly what we need.

DEFINITION 5.3   *A tournament* $T = (V, A)$ *is* transitive *if* $(u, v) \in A$ *and* $(v, w) \in A$ *implies* $(u, w) \in A$.

Of the two tournaments (d) and (e) of Figure 5.1 it is no accident that the first is transitive and the second is not. In fact, the uniqueness characterizes transitive tournaments.

THEOREM 5.3
A tournament T *is transitive if and only if there is a unique complete simple path.*

PROOF   Suppose T is transitive and has two complete simple paths. Then, there exists a pair $\{u, v\}$ of vertices such that u comes before v in one path and v before u in the other. But if there is a path from u to v, there is an arc (u, v) in transitive tournaments, and since (v, u) must also be an arc, we reach a contradiction.   ∎

Now, suppose there is a unique complete simple path $u_1, u_2, \ldots, u_n$ and let $i < j$ so that u_i comes before u_j. We first show that $(u_i, u_j) \in A$. From this, we can conclude that $(u_i, u_j) \in A$ implies $i < j$, since if $j > i$, then $(u_j, u_i) \in A$. Choose the smallest i for which $(u_i, u_j) \in A$ and the largest corresponding j. Then, if $i > 1$, we construct a new complete simple path as shown in Figure 5.4: $u_1, \ldots, u_{i-1}, u_{i+1}, \ldots, u_j, u_i, u_{j+1}, \ldots, u_n$. If $i = 1$, the new path begins with u_{i+1}.

This construction tells us that $i < j$ if and only if $(u_i, u_j) \in A$. The transitivity is now clear, since if $(u_i, u_j) \in A$ and $(u_j, u_k) \in A$, then $i < j$ and $j < k$, which implies $i < k$ and, hence, $(u_i, u_k) \in A$.

COROLLARY 5.4
A transitive tournament has a unique ranking based on the unique complete simple path.

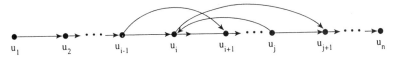

Figure 5.4: Construction of the new complete simple path.

The ranking given by the score sequence can also be established for a transitive tournament. For the transitive tournament of order 3 in Figure 5.1b, the score sequence is 2, 1, 0; for that of order 4 in Figure 5.1d, the score sequence is 3, 2, 1, 0.

COROLLARY 5.5
A transitive tournament has a unique ranking based on the score sequence which is the same as that given by the complete simple path.

PROOF The score sequence for such a tournament of order n is $(n-1)$, $(n-2), \ldots, 3, 2, 1, 0$. This may be shown by observing that the score of the last vertex u_n in the path $u_1, u_2, \ldots, u_n$ is 0; otherwise there would be another complete simple path. This last vertex is removed to obtain a tournament of order $n-1$ with path $u_1, u_2, \ldots, u_{n-1}$. Again, the score of u_{n-1} in this new tournament is 0, but in the previous tournament, it is therefore 1. ∎

Problems 5.3

1. Find the score sequence of the tournaments in Figure 5.5.

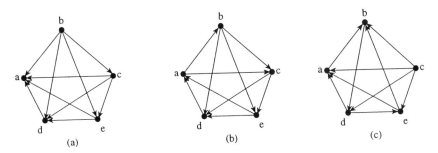

Figure 5.5: Three tournaments with five players.

2. (a) Which of the tournaments in Problem 1 is transitive? (b) Which has a unique winner? (c) Show that the winner in each case satisfies Proposition 5.1. (d) Find a complete simple path.

3. Show that the sum of all scores is $\frac{n(n-1)}{2}$ if a tournament has n players.

4. Show that all tournaments of order 4 are given in Figure 5.1. (Why can't 3, 3, 0, 0, be a score sequence?)

5. Show that two isomorphic tournaments have the same score sequence.

6. Let T be a tournament with score sequence 3, 2, 2, 2, 1. Show there is a complete simple path starting from any vertex. Is this always true for strongly connected tournaments? Imitate the proof of Proposition 3.2.

7. In a recent presidential primary election, there were five candidates B, C, H, K, T. A committee of three was to choose the candidate to be supported by the local party. The first member ranked them in the order B, C, H, K, T, the second K, T, B, C, H; and the third H, C, T, K, B. A preference digraph based on majority rule is as in Figure 5.6, which seemed to point to C as the winner. The chair of the committee preferred B and came up with the scheme: First choose between C and K, then between T and the winner of the first round, then between H and the winner of the second round, and then between B and the third round winner. Explain. Would this work in a transitive digraph? Would this work to make anyone the victor in a strongly connected tournament?

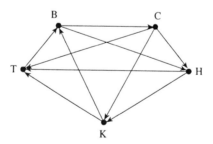

Figure 5.6: A presidential preference digraph.

8. Is a nontransitive tournament necessarily strongly connected? Prove or give a counterexample.

6

Planar Graphs

Although the diagrams of those graphs we have drawn have always been in
a plane, sometimes the edges cross each other. If the graph is isomorphic
to one where this doesn't occur, it is said to be **planar**. Thus, the graphs
in Figure 6.1 are planar, but the one in Figure 6.2 is not, but this is not so
easy to see yet.

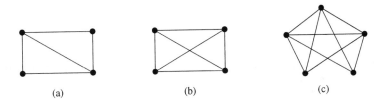

Figure 6.1: Some planar graphs.

Figure 6.2: A nonplanar graph.

A standard problem which needs the concept of planar graphs is the
houses and utility problem. A town has three houses and three utilities G,
W, and E. Is it possible to connect the three utilities to each of the houses
in such a way that the lines do not cross? There seems to be no obvious
way of constructing the lines in Figure 6.3; i.e., there seems to be no way
of connecting E to h_1 if the other connections are as shown.

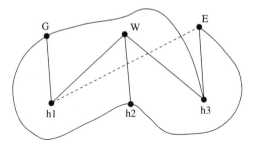

Figure 6.3: The house and utility problem.

This question of which graphs are planar is important in the construction of electrical circuits, in particular in the construction of computer chips.

6.1   Bipartite Graphs

Two types of graphs are important in the consideration of planar graphs. They are complete graphs and bipartite graphs. A **complete graph** on n vertices K_n is one which has the maximum possible number $n(n-1)/2$ of edges. The graph in Figure 6.2 is K_5; the one in Figure 6.1b is K_4.

DEFINITION 6.1   A graph $G = (V, E)$ is called bipartite *if $V = V_1 \cup V_2$, where V_1 and V_2 are disjoint, and for each $u \in V_1$ and $v \in V_2$, $\{u, v\} \in E$, but E has no other edges. If p = order of V_1 and g = order of V_2, then we denote by $K(p, q)$ the corresponding bipartite graph.*

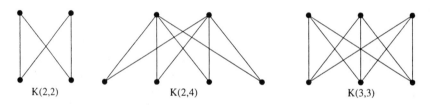

Figure 6.4: Some bipartite graphs.

Some of the bipartite graphs are clearly planar, for example, $K(2,2)$ and $K(2,4)$ in Figure 6.4, but $K(3,3)$ is the graph of our house and utility problem. In order to investigate this further, we consider the partitions of the plane associated with a planar graph. If the diagram of such a planar graph G is drawn in the plane, its edges partition the plane into disjoint regions.

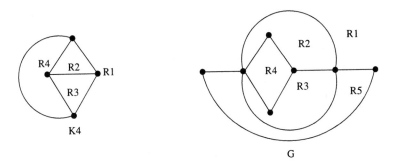

Figure 6.5: Partitioning the plane.

Although the regions themselves are not invariant under planar isomorphisms, the number of such regions is. If we let p = order of the graph, q = the number of edges, and r = the number of regions, then we find, for example in the case of K_4, that $p = 4$, $q = 6$, and $r = 4$. For G in Figure 6.5, we have $p = 7$, $q = 10$, and $r = 5$. For $K(2,4)$, we have $p = 6$, $q = 8$, and $r = 4$. Notice that in all of these cases, $p - q + r = 2$. Is this always the case?

6.2   A Necessary Condition for a Graph to Be Planar

Although we cannot give a simple test for establishing whether or not a graph is planar, we can obtain some formulas which hold for connected graphs.

THEOREM 6.1
Let G be a connected planar graph, and let p, q, and r be as above; then,
$p - q + r = 2$.

PROOF   The proof is by induction. It is clearly true for $p = 2$ since, then, $q = 1$ and $r = 1$, and is, in fact, true for all trees since $q = p - 1$ and $r = 1$ in this case. Now, let p be fixed and > 2, and suppose the result holds for all q such that $p - 1 \leq q \leq k$. We must show that it holds for $q - k + 1$. In this case, G is not a tree and, hence, must contain at least one cycle. Let e be an edge on this cycle and let $G_1 = G - \{e\}$. Then G_1 is connected and planar and has k edges and p vertices. If it has $r - 1$ regions, then $p - k + r - 1 = 2$ by the induction hypothesis. The edge partitions one of the regions into two parts. Hence, G has p vertices, $k + 1$ edges, and r regions. It satisfies

$$p - (k + 1) + r = p - k + r - 1 = 2.$$

∎

We could use this theorem to show that a graph is not planar by showing this relation among p, q, and r does not hold. Unfortunately, we cannot specify r for a nonplanar graph, so this approach doesn't work. However, the following theorem does work.

THEOREM 6.2

Let G be a connected planar graph of order $p \geq 3$ with q edges; then, $q \leq 3p - 6$.

PROOF The theorem holds for $p = 3$ since $q \leq 3$. Let $p \geq 4$ and let $R_1, R_2, \ldots, R_i, \ldots, R_r$ be the regions obtained from the partitioning of the plane by G. We denote by $n(R_i)$ the number of edges bounding R_i, which is always ≥ 3. Then, $n = \sum_{i=1}^{r} n(R_i) \geq 3r$. But each edge either lies on the boundary between two regions and therefore is counted twice in n, or it is not part of any boundary and is not counted. Hence, $n \leq 2q$. The two inequalities together give us $er \leq 2q$. We use this together with the result of Theorem 6.1 solved for $r, = 2 + q - p$, to conclude that $e(2 + q - p) \leq 2q$ or $q \leq 3p - 6$. ∎

With this theorem we can perhaps answer the questions about K_5 and $K(3,3)$. In the former case, we have $q = 10$ and $p = 5$, and since $15 - 6 = 9 \leq 10$, K_5 is not planar. However, the test fails for $K(3,3)$. Fortunately, we can get a slightly stronger result by the same methods.

THEOREM 6.3

Let G be a connected planar graph with no cycles of length $\leq c$; then, $q \leq c(p - 2)/(c - 2)$.

The proof is the same except that $n(R_i) \geq c$.

This result now shows that $K(3,3)$ is not planar and, therefore, the house and utility problem has no solution.

Problems 6.3

1. Show that the bipartite graphs $K(2,n)$ are planar for every n. Find p, q, and r.

2. For which values of m and n is $K(m,n)$ planar?

3. Show that all graphs of order 4 are planar.

4. In a marriage problem, a set of n men and n women must be matched consistent with an associated compatibility graph. This graph G is a subgraph of $K(n,n)$ with the same vertex set. The idea is to make the maximum possible numbers of matches. Let G be given by Figure 6.6.

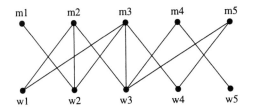

Figure 6.6: A graph for the marriage problem.

Use a breadth first search to obtain a spanning tree starting from w_1, say. Any simple path in this tree establishes a partial matching based on alternate edges. Start from other vertices as well as to try to maximize the number of matches. Vertices not on the simple path can be paired as well, if they lie on a branch of even length.

7

Graphs and Matrices

In order to store a graph or digraph in a computer, we need something other than the diagram or the formal definition. This something is the adjacency matrix, a matrix of 0's and 1's. The 1's correspond to the arcs of the digraph. Certain matrix operations will be seen to correspond to digraph concepts.

7.1 Adjacency and Reachability Matrices

We first give a formal definition and then shall see how to construct and use them.

DEFINITION 7.1 Let $D = (V, A)$ be digraph of order n, $V = \{u_1, u_2, \ldots, u_n\}$. The adjacency matrix $[a_{ij}]$ is an $n \times n$ matrix in which $a_{ij} = 1$ if $(u_i, u_j) \in A$ and $a_{ij} = 0$ if not. For a graph, the adjacency matrix is that of the associated symmetric digraph.

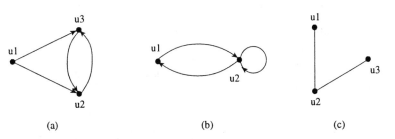

Figure 7.1: Graphs and digraphs for adjacency matrices.

The adjacency matrices for the graphs and digraphs of Figure 7.1 are

$$
\text{(a) } A = \begin{matrix} & u_1 & u_2 & u_3 \\ u_1 \\ u_2 \\ u_3 \end{matrix} \begin{pmatrix} 0 & 1 & 1 \\ 0 & 0 & 1 \\ 0 & 1 & 0 \end{pmatrix}, \quad \text{(b) } B = \begin{bmatrix} 0 & 1 \\ 1 & 1 \end{bmatrix}, \quad \text{(c) } C = \begin{bmatrix} 0 & 1 & 0 \\ 1 & 0 & 1 \\ 0 & 1 & 0 \end{bmatrix}
$$

If the vertices are listed next to the rows and columns as in a), the matrix has a 1 in the ith row and j the column if (u_i, u_j) is an arc. We shall generally denote the adjacency matrix by A; this will not now denote the set of arcs unless so specified.

THEOREM 7.1

Let D be a digraph (or graph) with adjacency matrix A; then, the i, j element of A^k gives the number of paths of length k in D from u_i to u_j.

For example, in Figure 7.1a, the adjacency matrix satisfies

$$
A^2 = \begin{bmatrix} 0 & 1 & 1 \\ 0 & 1 & 0 \\ 0 & 0 & 1 \end{bmatrix}, \quad A^3 = \begin{bmatrix} 0 & 1 & 1 \\ 0 & 0 & 1 \\ 0 & 1 & 0 \end{bmatrix}.
$$

Each 1 in A^2 comes from a path of length 2 and each 1 in A^3 comes from a path of length 3.

PROOF The proof is by induction on k. The result clearly holds for $k = 1$; we assume it to be true for $k = m$. Let u_i and $u_j \in V$ and suppose that there are p_{ij} paths of length $(m + 1)$ from u_i to u_j. Each such path ends in an arc, say (u_p, u_j). The path from u_i to u_p is of length m. By the induction hypothesis, there are $a_{ip}^{(m)}$ such paths. The number of paths from u_i to u_j is obtained by summing over all $a_{ip}^{(m)}$ such that there is an arc from some u_p to u_j, i.e., $p_{ij} = \sum_p a_{ip}^{(m)} a_{pj}$. But this is exactly the definition of matrix product, so $A^{m+1} = [p_{ij}]$. ∎

Another matrix associated with digraphs is the reachability matrix, which is useful for looking at questions of connectivity.

DEFINITION 7.2 *The reachability matrix $R = [r_{ij}]$ of a digraph has $r_{ij} = 1$ if there is a path from u_i to u_j, and 0 otherwise.*

In the digraph of Figure 7.1a, we have $R = \begin{bmatrix} 1 & 1 & 1 \\ 1 & 1 & 1 \\ 1 & 1 & 1 \end{bmatrix}$ since it is strongly connected, but for the digraphs in Figure 7.2, the matrices are

$$\text{(a)} \quad R = \begin{bmatrix} 1 & 1 & 1 \\ 0 & 1 & 1 \\ 0 & 0 & 1 \end{bmatrix} \qquad \text{(b)} \quad R = \begin{bmatrix} 1 & 1 & 1 \\ 1 & 1 & 1 \\ 0 & 0 & 1 \end{bmatrix}.$$

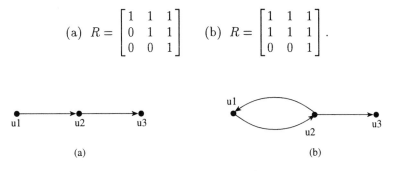

(a)   (b)

Figure 7.2: Digraphs for reachability matrices.

There should be some relation between a reachability and adjacency matrix since A^k gives us paths of length k. We first introduce a definition.

DEFINITION 7.3   Let $C = [c_{ij}]$ be a matrix. The Heaviside function applied to matrices is

$$H(C) = [H(c_{ij})],$$

where $H(x) = \begin{cases} 1 & x > 0 \\ 0 & x \leq 0 \end{cases}$.

For example, if $C = \begin{bmatrix} 0 & -2 & 3 \\ 0 & 1 & 1.5 \\ 0.5 & 0 & -1 \end{bmatrix}$, then $H(C) = \begin{bmatrix} 0 & 0 & 1 \\ 0 & 1 & 1 \\ 1 & 0 & 0 \end{bmatrix}$.

THEOREM 7.2
Let A be the adjacency matrix of a digraph D of order n, let

$$C = I + A + A^2 + \cdots + A^{n-1};$$

then $R = H(C)$ and has no zero elements if and only if D is strongly connected.

PROOF   If there is a path from u_i to u_j in D, there is a simple path; its length is $< n$. Hence, there must be a 1 or more in the i,j position of one of the matrices I, A, $A^2, \ldots, A^{n-1}$. Conversely, there must be a path if any one of these matrices has at least a 1 in the i,j position. There may be a number of paths between these two vertices; hence, C will have a positive number and $H(C)$ a 1 whenever there is at least one path.   ∎

If D is strongly connected, u_i and u_j can be any vertices and there is a path between them.

The square of the reachability matrix also has an important interpretation. Its diagonal elements give the number of vertices in the strongly connected component containing that vertex. In the digraph of Figure 7.2b, we find that

$$
R^2 = \begin{bmatrix} 1 & 1 & 1 \\ 1 & 1 & 1 \\ 0 & 0 & 1 \end{bmatrix}^2 = \begin{bmatrix} 2 & 2 & 3 \\ 2 & 2 & 3 \\ 0 & 0 & 1 \end{bmatrix}.
$$

Thus, the component containing u_1 has two elements, u_2 has two elements, and u_3 has one element.

Problems 7.1

1. Calculate the adjacency matrices for all tournaments of order 4.

2. Find the matrices A, R, and C for the digraphs of Figure 7.3. Show that $R = H(C)$.

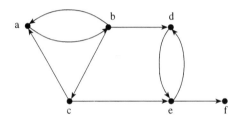

Figure 7.3: Example digraph.

3. Show that $R = H(I + A + A^2 + A^3)$ for a tournament with score sequence $(2, 2, 1, 1)$.

4. Find R^2 in the matrices of Problems 2 and 3 and check that the diagonal elements give the number of vertices in the components.

7.2 Eigenvalues of Adjacency Matrices

We first extend the definition of adjacency matrix to weighted digraphs.

DEFINITION 7.4   *The* adjacency matrix $A = [a_{ij}]$ *of a weighted digraph* $D = (V, A, w)$ *is given by* $a_{ij} = w(u_i, u_j)$.

The same definition applies to graphs as well. With this definition, any square matrix can be considered the adjacency matrix of a weighted digraph (provided we allow negative weights).

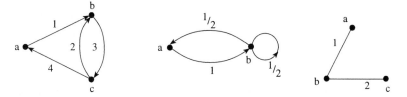

Figure 7.4: Some weighted digraphs and graphs for adjacency matrices.

The eigenvalues of a square matrix A are those real or complex numbers λ such that

$$\det(A - \lambda I) = 0$$

(see Appendix A.2).

a) $A = \begin{bmatrix} 0 & 1 & 0 \\ 0 & 0 & 3 \\ 4 & 2 & 0 \end{bmatrix}$ b) $A = \begin{bmatrix} 0 & 1 \\ \frac{1}{2} & \frac{1}{2} \end{bmatrix}$ c) $A = \begin{bmatrix} 0 & 1 & 0 \\ 1 & 0 & 2 \\ 0 & 2 & 0 \end{bmatrix}$

For example, in matrix b above, the eigenvalues are solutions to

$$\det\left(\begin{bmatrix} 0 & 1 \\ \frac{1}{2} & \frac{1}{2} \end{bmatrix} - \lambda \begin{bmatrix} 1 & 0 \\ 0 & 1 \end{bmatrix} \right) = \det \begin{bmatrix} -\lambda & 1 \\ \frac{1}{2} & \frac{1}{2} - \lambda \end{bmatrix} = \lambda^2 - \frac{1}{2}\lambda - \frac{1}{2} = 0$$

or $\lambda = -\frac{1}{2}, 1$. For an $n \times n$ matrix, the eigenvalues are zeros of an nth-degree polynomial.

For nonweighted digraphs, the eigenvalue of the adjacency matrix are sometimes simpler to find. For example, in a cycle of length 3, the adjacency matrix is

$$A = \begin{bmatrix} 0 & 1 & 0 \\ 0 & 0 & 1 \\ 1 & 0 & 0 \end{bmatrix}$$

and the eigenvalues must satisfy $\lambda^3 - 1 = 0$ or $\lambda = 1, e^{2\pi i/3}, e^{-2\pi i/3}$. For a transitive tournament with three vertices, the matrix is

$$A = \begin{bmatrix} 0 & 1 & 1 \\ 0 & 0 & 1 \\ 0 & 0 & 0 \end{bmatrix}$$

and all of the eigenvalues are zero as a result. In general, for cycles of length n, the eigenvalues must satisfy $\lambda^n = 1$. For other digraphs, we have the following theorem.

THEOREM 7.3
The number λ is an eigenvalue of the adjacency matrix of D if and only if it is an eigenvalue of the adjacency matrix of a strong component of D.

For example, this shows that the eigenvalues of a transitive tournament of any order are all zero. The eigenvalues of the tournament with score sequence $(2, 2, 2, 0)$ are 0, 1, and $e^{\pm 2\pi i/3}$.

PROOF Let D have components $K_1, K_2, \ldots, K_s$, with K_1 an element of the vertex basis of D^*. Denote by K_1' the subgraph of D generated by the vertices in $D - K_1$ (i.e., the subgraph including all arcs in D joining vertices in $D - K_1$). Let K_1 have vertices $u_1, u_2, \ldots, u_r$, and K_1' have vertices $u_{r+1}, \ldots, u_n$. Then, with the ordering we have,

$$A = \begin{bmatrix} A_1 & A_{12} \\ 0 & A_{22} \end{bmatrix}$$

since there are no arcs going to K_1 from K_1'. We also have

$$A - \lambda I = \begin{bmatrix} A_1 - \lambda I_1 & A_{12} \\ 0 & A_{22} - \lambda I_{22} \end{bmatrix}$$

and, hence, $\det(A - \lambda I) = \det(A_1 - \lambda I_1)\det(A_{22} - \lambda I_{22})$. Thus, each eigenvalue of A_1 is an eigenvalue of A and each eigenvalue of A is an eigenvalue of A_1 or A_{22}. The procedure can now be repeated with K_1' and A_{22} to find that

$$\det(A - \lambda I) = \det(A_1 - \lambda I_1)\det(A_2 - \lambda I_2)\cdots\det(A_s - \lambda I_s),$$

where $A_1, \ldots, A_s$ are the adjacency matrices of the components $K_1, \ldots, K_s$ respectively. ∎

Problems 7.2

1. Find all the eigenvalues for all tournaments of order 4. (You may have to approximate some.)

2. Find the digraphs or weighted digraphs of

$$A = \begin{bmatrix} 1 & 5 & 2 \\ 0 & 0 & -1 \\ -1 & 0 & -1 \end{bmatrix}, \quad B = \begin{bmatrix} 1 & 1 & 0 & 0 & 0 \\ 1 & 1 & 0 & 0 & 0 \\ 0 & 0 & 0 & 1 & 0 \\ 0 & 0 & 1 & 0 & 0 \\ 0 & 0 & 0 & 0 & 1 \end{bmatrix},$$

$$C = \begin{bmatrix} 1 & 2 & 3 & 4 \\ 2 & 0 & 1 & 2 \\ 0 & 0 & 1 & 1 \\ 0 & 0 & 1 & 0 \end{bmatrix}.$$

3. Does Theorem 7.3 hold for weighted digraphs as well? Why?

4. Find the eigenvalues of the matrices in Problem 2 by using Theorem 7.3.

7.3 Using Maple with Graphs

A number of symbolic manipulation computer programs have become available in the last few years. These programs enable one to solve algebraic and differential equations and do many other mathematical manipulations in a symbolic or exact form. They can also do numerical calculations and can plot curves and surfaces. These programs, which include Derive, Macsyma, Maple, and Mathematica, can do some of the operations associated with graphs and digraphs of the sort with which we have been concerned. In this section, we shall concentrate only on Maple and will give a few of the graph-theoretic procedures available in it.

Maple can be used to construct either graphs or digraphs. In either case, the command with(networks): must precede other commands associated with graphs. To construct a graph, you must list the vertex set and the edge set. For example, the graph shown in Figure 7.5 is defined by the command

$$G := \text{graph}(\{1, 2, 3\}, \{\{1, 2\}, \{1, 3\}\}):$$

That is, the graph G has vertex set $\{1, 2, 3\}$ and the two edges $\{1, 2\}$ and $\{1, 3\}$. Similarly, a digraph may be defined by specifying its vertex and arc sets. The digraph in Figure 7.5 would be defined by

$$H := \text{graph}(\{1, 2, 3\}, \{[1, 2], [2, 3], [1, 3]\}):$$

Notice that the arcs have square brackets, whereas the edges have curly brackets.

Maple will draw the diagram of the graph automatically if you use the command

$$\text{draw}(G);$$

You will see the graph as shown in Figure 7.5. If you try the same thing with the digraph, you again get a diagram, but the arcs will have no arrows to indicate direction. Nonetheless, the program will store the directions of the arcs and can use them in appropriate operations. The directions can

Figure 7.5: Illustrative examples of graphs and digraphs.

be found by using the command

$$\text{head}(\mathsf{H});$$

which will give the heads of the arrows; in this case, vertices 2, 3, and 3.

Some of the operations that we have done by hand can now be done automatically. For example, to construct a minimum weight spanning tree, we can use the command

$$\text{spantree}(\mathsf{G}) :$$

which, in our case, in a spanning tree with the fewest edges. This will generally be nondirected; if we want a directed spanning tree ending with a certain vertex, we use a different command. It is

$$\mathsf{T} := \text{shortpathtree}(\mathsf{G}, 1) :$$

and creates a digraph T which spans G and has a path from each vertex to vertex 1 (i.e., vertex 1 is a vertex contrabasis of T). The path between any two vertices in this spanning tree can now be found by the command

$$\text{path}([\mathsf{a}, \mathsf{b}], \mathsf{T});$$

Other useful commands are

$$\text{daughter}(\mathsf{v}, \mathsf{T});$$

which lists the vertices with paths from vertex v and

$$\text{ancestor}(\mathsf{v}, \mathsf{T});$$

which lists all the vertices with paths to vertex v.

There are many other commands in Maple for graphs, but they are oriented more to engineering applications than to those considered here. They may be found by using the help facility in Maple.

Problems 7.3

1. Use Maple to construct the two graphs in Figure 7.5. Use the draw command to check that they have the correct diagrams.

2. Find the components of the two graphs by using the component command.

3. Modify the first graph constructed in 1 by adding an edge between vertices u_3 and u_5. There is a command for doing this is Maple; find it in the help menu.

4. Use Maple to construct the second graph in Figure 4.8. Show it has no bridges by using the bicomponent command. Find a spanning tree and a directed spanning tree by using the commands given above.

5. Construct the digraph for a tournament with score sequence 2, 2, 1, 1. Use the daughter command and the ancestor command for a vertex of highest score. Give an interpretation of the meaning of the results in the context of a round robin tournament.

6. Use the isplanar command to check whether the graph of Problem 4 is planar or not.

7. The adjacency matrix of a graph or digraph may be found by using the adjacency command. Find the adjacency matrix of the digraph in Problem 5; then find the matrix corresponding to the first three powers. (You will need to load the linear algebra package by using with(linalg); the command multiply(A,A); gives the square of the matrix).

Part II

Digraphs and Probabilities: Markov Chains

As we saw in the previous part, there are many interesting questions that can be studied with the aid of digraphs. However, when it comes to flow models, these digraphs generally only give qualitative information. If we want to consider the quantitative dynamics of a problem, we have to associate a difference or differential equation with the flows. This will lead to either discrete or continuous compartmental models. In this part, we consider the discrete case which uses a difference equation approach. This leads to the theory of *Markov chains*, which initially seem complicated because they use the language of probability, but, in fact, use only some elementary properties of certain kinds of matrices.

8

Introduction to Markov Chains

In Chapter 7, we saw how weighted digraphs and adjacency matrices are related. In this chapter, we consider particular types of weights and matrices that are used with Markov chains, and their associated special terms, which differ from those used previously.

8.1 Relation to Digraphs

In this section, we give some definitions and a number of examples to try to clarify them.

DEFINITION 8.1 A stochastic digraph D is a weighted digraph in which the weights are positive numbers less than or equal to 1 such that the sum of the weights of the arcs leaving each vertex is 1.

In more concise terms:

If w_{ij} is the weight of the arc $(u_i, u_j) \in A$, such that $0 \leq w_{ij} \leq 1$ and $\sum_j w_{ij} = 1$, then D is a stochastic digraph.

Some examples are given in Figure 8.1.

The vertices in such a digraph are usually called *states* because they correspond to the outcomes of stochastic experiments. Many flow models can be interpreted as resulting from such experiments. This will become clear from the examples.

In example (1) of Figure 8.1, the experiment consists of tossing a coin with the two possible outcomes H and T. The four arcs, (H, H), (H, T), (T, H), (T, T) and (T, H) have weights corresponding to respectively the probability of getting a head given that a head was the previous outcome,

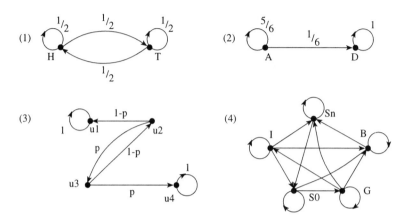

Figure 8.1: Some stochastic digraphs.

the probability of getting a tail given that a head was the previous outcome, etc.; all the arcs have weight $1/2$ for a fair coin.

In example (2), the experiment is Russian roulette as described in Chapter 1. There are two possible outcomes: A and D. The probability of going from state A to state D is $1/6$ and the probability of staying in state A is $5/6$.

Example (3) is the digraph of a random walk in which the walker staggers either to the left or to the right at each time step. Here, p is the probability of going to the right and $(1-p)$ is the probability of going to the left. Once you reach state u_1 or u_4, you stay there.

Example (4) comes from a model of a meadow ecosystem with insects (I), soils (So), grass (G), snakes (Sn), and birds (B). The arcs represent the flow of nutrients, say nitrates between the various states. The probability associated with each arc is that of the flow of a single molecule. The probability of (G, G) is that it remains in the state G in a particular week (or other time interval), while the probability of (So, G) is that the molecule moves from state So to G in that week.

8.2   More Definitions and Examples

The adjacency matrix of the stochastic digraph (often called a *transition matrix*) $P = [p_{ij}]$ has the properties that $0 \leq p_{ij} \leq 1$ and $\Sigma_j \, p_{ij} = 1$.

For the first three examples, we have

$$P_1 = \begin{bmatrix} \frac{1}{2} & \frac{1}{2} \\ \frac{1}{2} & \frac{1}{2} \end{bmatrix}, \quad P_2 = \begin{bmatrix} \frac{5}{6} & \frac{1}{6} \\ 0 & 1 \end{bmatrix}, \quad P_3 = \begin{bmatrix} 1 & 0 & 0 & 0 \\ 1-p & 0 & p & 0 \\ 0 & 1-p & 0 & p \\ 0 & 0 & 0 & 1 \end{bmatrix}.$$

What we have here in disguise is a *finite Markov chain*. Such a chain moves in a sequence of steps through the various states $u_1, u_2, \ldots, u_n$. The transition matrix gives the probability of going from state u_i to u_j in one step. Each row of P is a probability vector. The steps, called *trials*, correspond to one performance of the experiment.

The process is started with some *initial probability vector*

$$\mathbf{p}(0) = [p_1(0), p_2(0), \ldots, p_n(0)]$$

which may be either deterministic or stochastic. Usually, it will be of the form

$$\mathbf{p}(0) = [0, 0, \ldots, 0, 1, 0, \ldots, 0],$$

which indicates that it starts in the state with the 1, but it may also be started with some (other) random device. For example, we might be unsure where the molecule of nitrate is initially. Let

$$\mathbf{p}(t) = [p_1(t), p_2(t), \ldots, p_n(t)]$$

be the vector of probabilities where $p_i(t)$ is the probability of being in state u_i at time t (or trial t, or step t). Then,

$$\mathbf{p}(1) = \mathbf{p}(0)P,$$
$$\mathbf{p}(2) = \mathbf{p}(1)P,$$
$$\vdots$$
$$\mathbf{p}(t+1) = \mathbf{p}(t)P.$$

These may be combined into $\mathbf{p}(t) = \mathbf{p}(0)P^t$, $t = 0, 1, 2, \ldots$.

For example, in Russian roulette (Example 2), $\mathbf{p}(0) = [p_1(0), p_2(0)]$, the probability of being alive or dead at the start. If we start with, say, alive, then

$$\mathbf{p}(0) = [1, 0],$$

and $\mathbf{p}(1) = \left[\frac{5}{6}, \frac{1}{6}\right]$, $\mathbf{p}(2) = \left[\frac{25}{36}, \frac{11}{36}\right]$, $\mathbf{p}(3) = \left[\frac{125}{216}, \frac{91}{216}\right], \ldots$.

In the random walk example, if you start at u_2, then $\mathbf{p}(0) = [0, 1, 0, 0]$,

$$\mathbf{p}(1) = [1-p, 0, p, 0], \quad \mathbf{p}(2) = [0, p(1-p), 0, p^2].$$

Here are some additional examples.

$\begin{bmatrix} 1 & 0 \end{bmatrix} \begin{bmatrix} 5/6 & 1/6 \\ 0 & 1 \end{bmatrix} = \begin{bmatrix} 5/6, 1/6 \end{bmatrix}$; $\begin{bmatrix} 5/6, 1/6 \end{bmatrix} \begin{bmatrix} 5/6 & 1/6 \\ 0 & 1 \end{bmatrix} = \begin{bmatrix} 25/36, 11/36 \end{bmatrix}$

$p(0) \quad p^0 \qquad p(1) \qquad p(1) \qquad p^0 \qquad p(2)$

see p. 73

Example 5

In the city of M, there are three kinds of weather: good, bad, and awful. The probability of weather changes from one day to the next is given by the digraph in Figure 8.2.

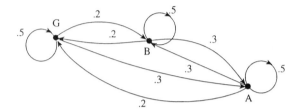

Figure 8.2: The digraph of the weather Markov chain.

Example 6

In the *gossip chain*, a rumor is spread from one person to another with a probability p of being passed on as it was heard and the probability $(1-p)$ of being passed on just the opposite of what was heard. Its digraph is shown in Figure 8.3.

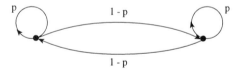

Figure 8.3: The digraph of the gossip Markov chain.

Example 7

A tall man in a certain society will have a tall son with probability .6, a medium son with probability .3, and a short son with probability .1. A medium-sized man will have the sons of the three sizes with probabilities .2, .5, and .3 respectively, and a short man will have sons of the three sizes with probabilities .2, .3, and .5 respectively.

Example 8

Three brands of beer, M, P, and S, are sold in a certain city. In a given month, 30% of brand M drinkers switched to brand P and 20% to brand

S. For brand P, 20% switched to M and 20% to S, whereas for brand S, nobody switched. If someone starts off with M, what is the probability he will stay with M in 3 months?

$$P(o) \; P^3 = [1,0,0] \begin{bmatrix} .5 & .3 & .2 \\ .2 & .6 & .2 \\ 0 & 0 & 1 \end{bmatrix}^3$$

Example 9
The probability that the party in power is to be returned to the presidency is .6. What portion of the time will each party (of two) be in power? The Markov chain which models this will have the same form as in Figure 8.3.

Example 10
In Leontief input–output analysis, an economy is split up into components each of which uses some of the output from other components. Suppose a carpenter, plumber, and electrician perform work on each other's houses. They put in the following hours in 1 week.

Output (work by)

		C	P	E
Input (house of)	C	8	4	24
	P	16	20	4
	E	16	16	12

How much should each be paid?

Notice that the matrix can be converted into a stochastic matrix by dividing each column by its sum (40) and then taking its transpose. The portion of the total wages going to each is the probability vector used. If all three start off with equal wages, i.e., if $p(0) = (1/3, 1/3, 1/3)$, then $p(1)$ is the relative cost to each. The wages are adjusted by making them equal to $p(1)$. In this way, we get a sequence of adjustments. If $p(t)$ has a limit as $t \rightarrow \infty$ (it does), then this is the amount each should be paid.

Example 11
A factory pollutes the Milwaukee River with cadmium. A given molecule washes out to Lake Michigan 99% of the time in a day. How long, on average, will it stay in the river?

$$.01 \bigcirc \xrightarrow{.99} \bigcirc 1 \qquad \begin{matrix} \mu_1 & \mu_2 \\ \mu_1 & .01 & .99 \\ \mu_2 & 0 & 1 \end{matrix}$$

$$E = \frac{1}{.99} = \frac{100}{99} \text{ days} \quad \text{(see p. 74)}$$

9

Classification of Markov Chains

We assume that we have a Markov chain with transition matrix P and stochastic digraph D, as described in the last chapter. The digraph can be assumed to be weakly connected since, otherwise, the chain can be split into several noninteracting parts.

9.1 Definitions

Two types of chains will be considered: *absorbing* and *regular* chains. An example of an absorbing chain is the Russian roulette chain in which D absorbs everything. Another example is the random walk. The weather chain or the Leontief input–output chain are both regular.

DEFINITION 9.1 *A state u_i is* absorbing *if its digraph has no arcs leaving it for other states. There may be a loop (u_i, u_i) (in fact, there must be and it must have weight 1).* $= \text{sink}$

DEFINITION 9.2 *A Markov chain* with digraph D is absorbing *if it contains at least one absorbing state and there is a path from each nonabsorbing state to some absorbing state.*

If the digraph is strongly connected, the Markov chain has no absorbing states, since there is a positive probability of leaving such state.

DEFINITION 9.3 *A* regular Markov chain *is one where the digraph D is strongly connected and there is a k such that any two vertices (which may be the same) can be joined by a path of exactly length k.*

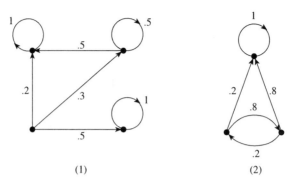

(1)   (2)

Figure 9.1: Digraphs of absorbing Markov chains.

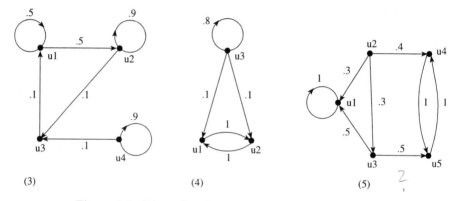

(3)   (4)   (5)

Figure 9.2: Digraphs of non-absorbing Markov chains.

Some digraphs which are not those of regular Markov chains are (1)–(5) in Figures 9.1 and 9.2 as well as (9) and (10) in Figure 9.4. Notice the similarity between (4) and (8). In (9) of Figure 9.4, the digraph is strongly connected, but the Markov chain is not regular since there is no path of an even length from u_1 to u_2 and no path of odd length from u_1 to u_1. The same is true for (10); the paths from u_1 to u_2 are of length 1, 4, 7, 11,..., whereas those from u_1 to u_3 are of length 2, 5, 8,.... However, by adding one more condition, we can obtain a sufficient condition for regularity.

THEOREM 9.1
Each strongly connected stochastic digraph with at least one loop is regular.

PROOF Since the digraph is strongly connected, there is a complete closed path. Thus, there is a path between any pair of vertices which

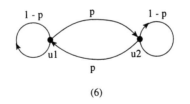

(6)

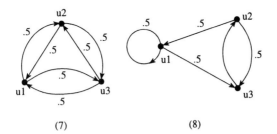

(7) (8)

Figure 9.3: Some digraphs of regular Markov chains.

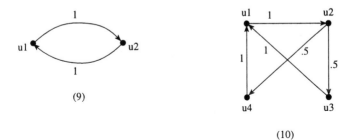

(9)

(10)

Figure 9.4: Nonabsorbing and nonregular Markov chains.

passes through the vertex with the loop, of length less than or equal the order of the digraph. We then form a new path which goes around the loop sufficiently often to make the length of the path equal to the order. ∎

The prototype of an absorbing chain is the Russian roulette of Figure 8.1(2). Its transition matrix raised to the nth power is

$$P^n = \begin{bmatrix} \frac{5}{6} & \frac{1}{6} \\ 0 & 1 \end{bmatrix}^n = \begin{bmatrix} \left(\frac{5}{6}\right)^n & 1 - \left(\frac{5}{6}\right)^n \\ 0 & 1 \end{bmatrix}.$$

As $n \to \infty$, this converges to

$$P^\infty = \begin{bmatrix} 0 & 1 \\ 0 & 1 \end{bmatrix}.$$

The expected number of trials before absorption is the inverse of $\frac{1}{6}$, i.e., 6. (A general method to determine this will be given later.)

The prototype of a regular chain is the *presidential power chain*(or the gossip chain) with transition matrix

$$P = \begin{bmatrix} 1-p & p \\ p & 1-p \end{bmatrix}.$$

Its digraph is given in Figure 9.3(6). Now, however, the powers of P are harder to calculate, e.g.,

$$P^2 = \begin{bmatrix} (1-p)^2 + p^2 & 2p(1-p) \\ 2p(1-p) & (1-p)^2 + p^2 \end{bmatrix},$$

$$P^3 = \begin{bmatrix} (1-p)^3 + 3p^2(1-p) & 3p(1-p)^2 + p^3 \\ 3p(1-p)^2 + p^3 & (1-p)^3 + 3p^2(1-p) \end{bmatrix}.$$

For P^n, we use the binomial theorem

$$(1 - p + p)^n = \sum_{k=0}^{n} (n!/k!(n-k)!)(1-p)^k p^{n-k}$$

and can show (by induction) that if

$$P^n = \begin{bmatrix} p_{11}(n) & p_{12}(n) \\ p_{21}(n) & p_{22}(n) \end{bmatrix},$$

then

$$p_{11}(n) = p_{22}(n) = (1-p)^n + (n!/(n-2)!2!)(1-p)^{n-2}p^2$$
$$+ (n!/(n-4)!4!)(1-p)^{n-4}p^4 + \cdots$$
$$= \text{even terms in the binomial expansion,}$$

and

$$p_{12}(n) = p_{21}(n) = (n!/(n-1)!1!)(1-p)^{n-1}p$$
$$+ (n!/(n-3)!3!)(1-p)^{n-3}p^3 + \cdots$$
$$= \text{odd terms in expansion.}$$

An alternative approach uses the eigenvalues and eigenvectors of the transition matrix P to find the powers more easily. The eigenvalues are obtained from the equation

$$\det(P - \lambda I) = (1 - p - \lambda)^2 - p^2 = (1 - \lambda - 2p)(1 - \lambda) = 0$$

and, hence, are

$$\lambda_1 = 1, \quad \lambda_2 = 1 - 2p.$$

The corresponding eigenvectors are

$$\mathbf{k}_1 = 2^{-\frac{1}{2}} \begin{bmatrix} 1 \\ 1 \end{bmatrix}, \quad \mathbf{k}_2 = 2^{-\frac{1}{2}} \begin{bmatrix} 1 \\ -1 \end{bmatrix};$$

the two together satisfy

$$P\mathbf{k} = 1\mathbf{k}_1, \quad P\mathbf{k}_2 = (1 - 2p)\mathbf{k}_2.$$

In matrix form, this is

$$P[\mathbf{k}_1, \mathbf{k}_2] = [\mathbf{k}_1, \mathbf{k}_2] \begin{bmatrix} 1 & 0 \\ 0 & 1 - 2p \end{bmatrix}$$

or

$$P = [\mathbf{k}_1, \mathbf{k}_2] \begin{bmatrix} 1 & 0 \\ 0 & 1 - 2p \end{bmatrix} [\mathbf{k}_1, \mathbf{k}_2]^{-1}.$$

From this, it is easy to find the powers of P as

$$P^n = [\mathbf{k}_1, \mathbf{k}_2] \begin{bmatrix} 1 & 0 \\ 0 & 1 - 2p \end{bmatrix}^n [\mathbf{k}_1, \mathbf{k}_2]^{-1} \longrightarrow [\mathbf{k}_1, \mathbf{k}_2] \begin{bmatrix} 1 & 0 \\ 0 & 0 \end{bmatrix} [\mathbf{k}_1, \mathbf{k}_2]^{-1}.$$

Since $[\mathbf{k}_1, \mathbf{k}_2]^{-1} = [\mathbf{k}_1, \mathbf{k}_2]$, we conclude that

$$P^n \longrightarrow [\mathbf{k}_1, \mathbf{k}_2] \begin{bmatrix} 1 & 0 \\ 0 & 0 \end{bmatrix} [\mathbf{k}_1, \mathbf{k}_2] = \begin{bmatrix} \frac{1}{2} & \frac{1}{2} \\ \frac{1}{2} & \frac{1}{2} \end{bmatrix} = 2^{-\frac{1}{2}} [\mathbf{k}_1, \mathbf{k}_2].$$

Problems 9.1

1. Which of the stochastic digraphs in Figure 9.5 are regular Markov chains, which are absorbing Markov chains, and which are neither?

2. Which of the following stochastic matrices are regular Markov chains, absorbing Markov chains, and neither?

   (a) $P = \begin{bmatrix} 1 & 0 \\ \frac{1}{5} & \frac{4}{5} \end{bmatrix}$, (b) $P = \begin{bmatrix} 0 & 1 \\ \frac{1}{5} & \frac{4}{5} \end{bmatrix}$, (c) $P = \begin{bmatrix} \frac{2}{3} & \frac{1}{3} \\ \frac{2}{3} & \frac{1}{3} \end{bmatrix}$,

   (d) $P = \begin{bmatrix} 0 & 0 & 1 \\ \frac{1}{3} & \frac{1}{3} & \frac{1}{3} \\ 0 & 0 & 1 \end{bmatrix}$, (e) $P = \begin{bmatrix} \frac{1}{2} & \frac{1}{2} & 0 \\ 0 & \frac{1}{2} & \frac{1}{2} \\ \frac{1}{2} & \frac{1}{2} & 0 \end{bmatrix}$.

3. Find $\lim_{n \to \infty} P^n$ for (a), (b) and (c) in Problem 2 by using the approach based on eigenvalues.

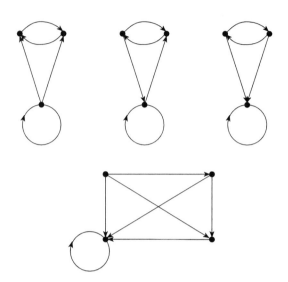

Figure 9.5: Stochastic digraphs for Problem 1 (without weights).

9.2   Condensation of a Stochastic Digraph

In Chapter 4, we saw that a digraph $D = (V, A)$ has a condensation $D^* = (V^*, A^*)$ whose vertex set V^* is the set of all strong components of D. We saw that D^* is acyclic and has a vertex basis consisting of all components with no incoming arcs. This same approach allows us to construct a condensation of a stochastic digraph and, hence, a reduction of a Markov chain.

Let $V^* = \{K_1, K_2, \ldots, K_n\}$ be the vertex set with vertex basis B^*. We must assign a probability to each arc $(K_i, K_j) \in A^*$. We do so by averaging all of the probabilities of arcs in A going from K_i to K_j; that is,

$$p(K_i, K_j) = \sum_{\substack{v_k \in K_i \\ v_m \in K_j}} p(v_k, v_m)/\#(K_i), \qquad (9.1)$$

where $\#(K_i)$ is the number of vertices in the component K_i. In particular, the loop probability is given by

$$p(K_i, K_i) = \sum_{v_k, v_j \in K_i} p(v_k, v_j)/\#(K_i).$$

Clearly, each $p(K_i, K_j)$ is between 0 and 1 since it is the average of numbers between 0 and 1. Furthermore, we have

$$\sum_{j=1}^{n} p(K_i, K_j) = 1$$

since

$$\sum_{j=1}^{n} p(K_i, K_j) = \frac{1}{\#(K_i)} \sum_{j=1}^{m} \sum_{\substack{v_k \in K_i \\ v_m \in K_j}} p(v_k, v_m)$$

$$= \frac{1}{\#(K_i)} \sum_{v_k \in K_i} \sum_{m=1}^{N} p(v_k, v_m) = \frac{1}{\#(K_i)} \sum_{v_K \in K_i} 1$$

$$= 1.$$

To illustrate this, we consider the stochastic digraph in Figure 9.6.

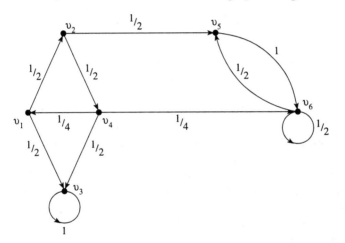

Figure 9.6: A stochastic digraph to illustrate the condensation.

The strongly connected components are $K_1 = \{v_1, v_2, v_4\}$, $K_2 = \{v_3\}$, and $K_3 = \{v_5, v_6\}$. The condensation stochastic digraph has probabilities

$$p(K_1, K_1) = \frac{1}{3}(p(v_1, v_2) + p(v_1, v_4) + p(v_2, v_4) + p(v_2, v_1)$$

$$+ p(v_4, v_1) + p(v_4, v_2))p(v_1, v_1) + p(v_2, v_4) + p(v_4, v_4)$$

$$= \frac{1}{3}\left(\frac{1}{2} + 0 + \frac{1}{2} + 0 + \frac{1}{4} + 0 + 0 + 0 + 0\right) = \frac{5}{12}.$$

$$p(K_1, K_2) = \frac{1}{3}\left(\frac{1}{2} + \frac{1}{2}\right) = \frac{1}{3},$$

$$p(K_1, K_3) = \frac{1}{3}\left(\frac{1}{2} + \frac{1}{4}\right) = \frac{1}{4},$$

$$p(K_2, K_2) = 1,$$

$$p(K_2, K_3) = p(K_2, K_1) = 0,$$

$$p(K_3, K_2) = p(K_3, K_1) = 0,$$

$$p(K_3, K_3) = \frac{1}{2}\left(1 + \frac{1}{2} + \frac{1}{2}\right) = 1;$$

thus, the condensation stochastic digraph is as in Figure 9.7.

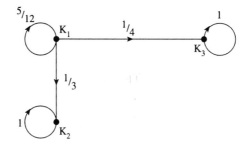

Figure 9.7: The condensation of the digraph of Figure 9.6.

Notice that this gives an absorbing Markov chain with absorbing states K_2 and K_3. These constitute the *vertex contrabasis* of D^* consisting of all vertices with no outgoing arcs. Since the digraph of D^* is acyclic, it is always possible to find a path from any vertex to an element of the vertex contrabasis, by the same argument as for the vertex basis. Thus, we have proved in the original matrix P that the submatrices P_{22} and P_{33} are each transition matrices of regular Markov chains. P_{22} is trivial and clearly $P_{22}^n \to [1]$ as $n \to \infty$. In the case of P_{33}, we find that $P_{33}^n \to W_3^\infty$ as $n \to \infty$, where

$$W_3^\infty = \begin{bmatrix} \frac{1}{2} & \frac{2}{3} \\ \frac{1}{2} & \frac{2}{3} \end{bmatrix}.$$

Hence, we have

$$P^n = \begin{bmatrix} P_{11}^n & P_{12}^{(n)} & P_{13}^{(n)} \\ 0 & P_{22}^n & 0 \\ 0 & 0 & P_{33}^n \end{bmatrix} \longrightarrow \begin{bmatrix} 0 & P_{12}^{(\infty)} & P_{13}^{(\infty)} \\ 0 & 1 & 0 \\ 0 & 0 & W_3^\infty \end{bmatrix}$$

and we conclude that everything must be absorbed in one of the two components K_2 and K_3 eventually.

The arguments in this example are general and lead to the following proposition.

PROPOSITION 9.2
Let D be a stochastic digraph and let D^* be its condensation with probabilities given by (9.1). Then, D^* is a stochastic digraph of an absorbing Markov chain.

We return to the example and compare the transition matrices of D and of D^*. In the former, we have

$$
P = \begin{array}{c} v_1 \\ v_2 \\ v_4 \\ v_3 \\ v_4 \\ v_6 \end{array}
\begin{pmatrix}
0 & \frac{1}{2} & 0 & \frac{1}{2} & 0 & 0 \\
0 & 0 & \frac{1}{2} & 0 & \frac{1}{2} & 0 \\
\frac{1}{4} & 0 & 0 & \frac{1}{2} & 0 & \frac{1}{4} \\
0 & 0 & 0 & 1 & 0 & 0 \\
0 & 0 & 0 & 0 & 0 & 1 \\
0 & 0 & 0 & 0 & \frac{1}{2} & \frac{1}{2}
\end{pmatrix}
= \begin{bmatrix}
P_{11} & P_{12} & P_{13} \\
0 & P_{22} & 0 \\
0 & 0 & P_{33}
\end{bmatrix}.
$$

which is in block triangular form. The matrix for D^* is

$$
P^* = \begin{bmatrix}
\frac{5}{12} & \frac{1}{3} & \frac{1}{4} \\
0 & 1 & 0 \\
0 & 0 & 1
\end{bmatrix},
$$

which has the same structure as the block triangular matrix of P. Since P^* is the matrix of an absorbing Markov chain (although not in standard form), we can find

$$
\lim_{n \to \infty} (P^*)^n = \begin{bmatrix}
0 & \frac{4}{7} & \frac{3}{7} \\
0 & 1 & 0 \\
0 & 0 & 1
\end{bmatrix}
$$

Problems 9.2

1. Find the condensation stochastic digraph of each of the stochastic digraphs in Figure 9.8.

(a) $K_1 = \{u_1\}$; $K_2 = \{u_2, u_3\}$

$p(K_1 K_1) = (1)(p u_{12}) = (1)(\frac{1}{2}) = \frac{1}{2}$ not right

$p(K_2 K_2) = \frac{1}{2}(p u_2 u_3) + p(u_3 u_2) = \frac{1}{2}(1+1) = 1$

$p(K_1 K_2) = \frac{1}{2}(p u_1 u_2) + p(u_1 u_3) = \frac{1}{2}(\frac{1}{4} + \frac{1}{4}) = \frac{1}{2}$

$G \xrightarrow{\frac{1}{2}} 1$

$\frac{1}{2}$

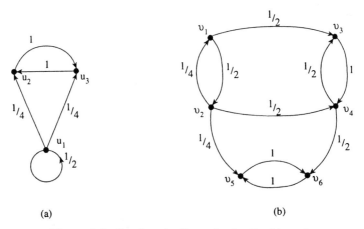

(a)   (b)

Figure 9.8: Stochastic digraphs for Problem 1.

2. Find the transition matrix P of each of the digraphs in 1, and find
   what happens to P^n as $n \to \infty$.

a) $P^* = \begin{bmatrix} 1/2 & 1/2 \\ 0 & 1 \end{bmatrix}$ $P^2 = \begin{bmatrix} 1/4 & 3/4 \\ 0 & 1 \end{bmatrix}$ $P^3 = \begin{bmatrix} 1/8 & 7/8 \\ 0 & 1 \end{bmatrix} \Rightarrow \begin{bmatrix} 0 & 1 \\ 0 & 1 \end{bmatrix}$

everything winds up in K_2

10

Regular Markov Chains

Let P be the transition matrix of a regular Markov Chain. Since by the definition of regular, there is a k such that any two vertices are joined by a path of length k, it follows that P^k has all positive elements. Recall that the powers of the adjacency matrix of a nonweighted digraph, A^k, count the number of paths from the vertex u_i to u_j. The same argument works for P^k.

10.1 Theory of Regular Chains

Example

In the regular chain given by the digraph in Figure 10.1 the matrix is

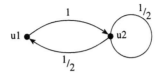

Figure 10.1: A simple regular chain.

$$P = \begin{bmatrix} 0 & 1 \\ .5 & .5 \end{bmatrix},$$

but since all pairs of vertices can be joined by a path of length 2, we find that

$$P^2 = \begin{bmatrix} .5 & .5 \\ .25 & .75 \end{bmatrix}$$

has all positive elements as expected.

In most of the examples of Chapter 8, we are interested in the behavior of P^k as $k \to \infty$. This will enable us to predict the ultimate proportion of the three kinds of beer in Example 8, the salary of each worker in Example 10, the proportion of tall, medium, and short men in Example 7, and the proportion of time each party will be in power in Example 9.

In the Leontief input–output (Example 10 of Section 8.2), we have

$$P = \begin{bmatrix} \frac{1}{5} & \frac{2}{5} & \frac{2}{5} \\ \frac{1}{10} & \frac{1}{2} & \frac{2}{5} \\ \frac{3}{5} & \frac{1}{10} & \frac{3}{10} \end{bmatrix}$$

and, therefore, it is clearly regular. Powers of P are

$$P^2 = \begin{bmatrix} .32 & .32 & .36 \\ .31 & .33 & .36 \\ .32 & .32 & .37 \end{bmatrix},$$

$$P^4 = \begin{bmatrix} .3132 & .3232 & .3636 \\ .3131 & .3232 & .3636 \\ .3131 & .3232 & .3637 \end{bmatrix},$$

$$P^{16} = \begin{bmatrix} .3131317 & .3232323 & .3636364 \\ .3131317 & .3232323 & .3636364 \\ .3131317 & .3232323 & .3636364 \end{bmatrix},$$

from which it appears that P^t approaches a positive matrix with identical rows:

$$\lim_{t \to \infty} P^t = \begin{bmatrix} \frac{31}{99} & \frac{32}{99} & \frac{36}{99} \\ \frac{31}{99} & \frac{32}{99} & \frac{36}{99} \\ \frac{31}{99} & \frac{32}{99} & \frac{36}{99} \end{bmatrix} = \begin{bmatrix} w_1 & w_2 & w_3 \\ w_1 & w_2 & w_3 \\ w_1 & w_2 & w_3 \end{bmatrix} = W.$$

This turns out to be the case for all regular Markov chains.

THEOREM 10.1

Let P be the transition matrix of a regular Markov chain. Then,

$$P^t \to W \text{ as } t \to \infty,$$

where W is a positive stochastic matrix, all of whose rows are the same vector $\mathbf{w}$, all of whose components are positive.

PROOF We first suppose P has all positive elements. We start off with a column vector $\mathbf{x}_0$ and multiply P by it on the right; denote the product by $\mathbf{x}_1 = P\mathbf{x}_0$. Let $m_0 = \min \mathbf{x}_0$, and $M_0 = \max \mathbf{x}_0$, the smallest and largest components of $\mathbf{x}_0$, respectively, and, similarly, m_1 and M_1 those of $\mathbf{x}_1$. Now, each component of $\mathbf{x}_1$, say $\mathbf{x}_{1i}$, is a weighted average of the components of $\mathbf{x}_0$ and, hence, satisfies

$$m_0 \leq x_{li} \leq M_0$$

and, in particular,

$$m_0 \leq m_1 \leq M_1 \leq M_0.$$

We now replace all elements of $\mathbf{x}_0$ except m_0 by M_0; the same inequality holds with the new $\mathbf{x}_{1i}$, and, in fact, each is a weighted average of m_0 and M_0. So the old M_1 is no larger than this weighted average, $M_1 \leq pM_0 + qm_0$, $p + q = 1$, $q > 0$. By a similar argument, $m_1 \geq p'm_0 + q'M_0$. The difference between the two satisfies

$$M_1 - m_1 \leq (1 - q - q')(M_0 - m_0). \tag{10.1}$$

We then repeat the argument with $\mathbf{x}_t = P^t\mathbf{x}_0$ to get

$$M_t - m_t \leq (1 - q - q')^t(M_0 - m_0). \tag{10.2}$$

Since $q > 0$, $M_t - m_t \to 0$ as $t \to \infty$; furthermore, $\{m_t\}$ is monotonically nondecreasing and bounded and, thus, converges as $t \to \infty$. Also, $\{M_t\}$ converges to the same value, as do all the elements of $\mathbf{x}_t$.

If P has some zero elements, we use P^k instead and deduce that $M_{kt} - m_{kt}$ satisfies (10.2) and, hence, converges to 0 as $t \to \infty$. We then take $t = ks + r$ and observe that

$$M_{ks+r} - m_{ks+r} \leq M_{ks} - m_{ks}$$

even when P has some zero elements.

To obtain the final conclusion about W, we take $\mathbf{x}_0$ to be the vector with 1 in the ith position and zeros elsewhere. Then, $\mathbf{x}_t$ converges to the ith column of W, all of whose components are the same.

Clearly, all the components of $\mathbf{w}$ are positive since $\mathbf{w} = \mathbf{w}P^k$ and P^k is positive. ∎

COROLLARY 10.2
Let $\mathbf{p}_0$ *be an initial probability vector; then,*

$$\mathbf{p}_t = \mathbf{p}_0 \, P^t \to \mathbf{p}_0 \, W = \mathbf{w}.$$

PROOF

$$\mathbf{p}_0 \, W = [p_{10}, \dots, p_{n0}] \begin{bmatrix} w_1 & w_2 & \cdots & w_n \\ w_1 & w_2 & \cdots & w_n \\ \cdots & \cdots & & \cdots \\ w_1 & w_2 & \cdots & w_n \end{bmatrix} = \left[w_1 \sum_i p_{i0} \cdots w_n \sum_i p_{i0} \right]$$

$$= \mathbf{w}.$$

∎

10.2   Fixed-Point Probability Vector

As an application of Corollary 10.2, we consider the political power chain of Example 9 of Chapter 8. In this case, it is easy to see that each party will ultimately be in power half the time, since

$$P = \begin{bmatrix} .6 & .4 \\ .4 & .6 \end{bmatrix}$$

and, hence, P^t converges to

$$W = \begin{bmatrix} .5 & .5 \\ .5 & .5 \end{bmatrix},$$

as we can see by finding successively higher powers of P. Hence, $\mathbf{w} = [.5 \; .5]$. Taking the limit is the hard way to find $\mathbf{w}$. A much easier way is to observe that $\mathbf{w}$ is a left eigenvector of P corresponding to the eigenvalue 1. Indeed, let $\mathbf{v}$ be such an eigenvector; then,

$$\mathbf{v} = \mathbf{v}P = \mathbf{v}P^t \to \mathbf{v}W = \mathbf{w},$$

i.e., $\mathbf{v} = \mathbf{w}$. But how do we know there is a left eigenvector of P of eigenvalue 1? (There is always a right eigenvector of eigenvalue 1 since each row sums to 1.)

We now return to the Leontief Example 10 of Chapter 8 and try to find the eigenvector. It satisfies

$$\mathbf{w}P = \mathbf{w} \quad \text{or} \quad \mathbf{w}(P - I) = \mathbf{0},$$

or

$$[w_1 \; w_2 \; w_3] \begin{bmatrix} -.8 & .4 & .4 \\ .1 & -.5 & .4 \\ .6 & .1 & -.7 \end{bmatrix} = [0 \; 0 \; 0].$$

Since $\mathbf{w}$ is a probability vector and the matrix is singular, we may replace the first column of the matrix by 1's and the right side by $[1\ 0\ 0]$. We also first multiply both sides by 10 to get

$$[w_1\ w_2\ w_3] \begin{bmatrix} 1 & 4 & 4 \\ 1 & -5 & 4 \\ 1 & 1 & -7 \end{bmatrix} = [1\ 0\ 0].$$

We then use Gaussian elimination on the transpose of the augmented matrix. This gives us $w_3 = \frac{4}{11}$, $w_2 = \frac{1}{9}\left(4 - 3 \cdot \frac{4}{11}\right) = \frac{33}{99}$, and $w_1 = 1 - \frac{32}{99} - \frac{4}{11} = \frac{31}{99}$.

The vector $\mathbf{w}$ is, for obvious reasons, denoted the fixed-point probability vector.

Problems 10.2

1. Find the fixed-point probability vector for the following examples of Chapter 8:

   a. The weather chain (Example 5),

   b. The gossip chain (Example 6),

   c. Tall man heredity (Example 7).

2. The transition matrix of a 2-state regular chain in its most general form is

$$P = \begin{bmatrix} 1-a & a \\ b & 1-b \end{bmatrix},$$

   where $0 < a \le 1$ and $0 < b \le 1$ $(ab \ne 1)$. Find all eigenvalues and eigenvectors of P and write P in the form

$$P = K\Lambda K^{-1},$$

   where K is the (right) eigenvector matrix and Λ is the eigenvalue matrix. Use this to find $W = \lim_{t \to \infty} P^t$ directly.

3. In a closed aquatic ecosystem, a molecule of a nitrate may reside in any of three different states: in solution (S), in the phytoplankton (P), or in zooplankton (Z). The probability of the molecule going from S to P in a given day is .40 and the probability of it going from P to Z is .20, whereas the probability of going from either P or Z to S is .10.

   Sketch the stochastic digraph of the associated Markov chain and show that it is regular. Find the distribution of the nitrate in the three states after an extended period of time.

10.3   Influence Digraph

A group of n people $u_1, u_2, \ldots, u_n$ has to make a decision. In a legislature, they might decide how much money to spend on a particular project. Each member has a degree of influence over every other member (which may be zero). We assume that the influence of u_i on u_j is a number α_{ij} between 0 and 1 such that the total influence by each member is $\sum\limits_{i=1}^{n} \alpha_{ij} = 1$. At each step (say vote), each member changes his position based on the influence other members have on him/her. If, at step t, each member u_i wants to spend $a_i(t)$, at step $t+1$ he will want to spend $a_i(t+1) = \sum\limits_{k=1}^{n} \alpha_{ki} a_k(t)$.

The influence digraph will be a weighted digraph. An example is shown in Figure 10.2.

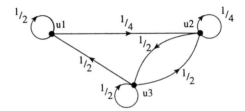

Figure 10.2: An influence digraph.

This is *not* a Markov chain, but would be if the arrows were turned around. We have

$$\mathbf{a}(t+1) = \mathbf{a}(t)A,$$

where A is the transition matrix of the weighted Markov chain obtained by turning the arcs around. By taking transposes, we have

$$\mathbf{a}^T(t+1) = A^T \mathbf{a}^T(t) = P\mathbf{a}^T(t),$$

where P is a stochastic matrix. Thus, any result that we have for Markov chain matrices applies to P. If, for example, the chain is regular, then

$$P^t \to W$$

and, hence,

$$P^t \mathbf{a}^T(0) \to W\mathbf{a}^T(0) = \begin{bmatrix} \sum w_i a_i(0) \\ \sum w_i a_i(0) \\ \vdots \\ \sum w_i a_i(0) \end{bmatrix}$$

i.e., the members of the group attain a final shared opinion.

If there are dominant (or stubborn) individuals who are not influenced by anybody but do exert some influence on others (for example, if the influence digraph is as in Figure 10.3),

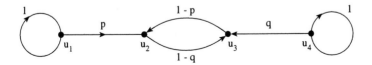

Figure 10.3: Another influence digraph.

then A^T is the matrix of an absorbing Markov chain; in the example,

$$A = \begin{array}{c} u_1 \\ u_4 \\ u_2 \\ u_3 \end{array} \begin{pmatrix} 1 & 0 & p & 0 \\ 0 & 1 & 0 & q \\ 0 & 0 & 0 & 1-q \\ 0 & 0 & 1-p & 0 \end{pmatrix}.$$

Problems 10.3

1. What happens in the long run in the example of Figure 10.2? Use the results for a regular Markov chain.

2. Do the same for the example of Figure 10.3.

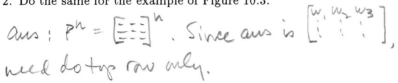

11

Absorbing Markov Chains

11.1 An Example

The prototypes of the absorbing chains are the Russian roulette and random walk chains. In the latter, the digraph (for four vertices) is given by Figure 11.1 and the transition matrix is

$$
P = \begin{array}{c} \\ u_1 \\ u_4 \\ u_2 \\ u_3 \end{array}
\begin{array}{cccc}
u_1 & u_4 & u_2 & u_3 \\
\end{array}
\left(
\begin{array}{cccc}
1 & 0 & 0 & 0 \\
0 & 1 & 0 & 0 \\
1-p & 0 & 0 & p \\
0 & p & 1-p & 0
\end{array}
\right)
= \left[\begin{array}{cc} I & O \\ R & Q \end{array} \right].
$$

The absorbing states are listed first, followed by the others. The form of P is shown by the partitioned matrices and is perfectly general for absorbing chains. Some natural questions that arise are the following:

Given that the chain starts in some nonabsorbing state u_i, the following questions must be addressed: (i) What is the probability of entering a given absorbing state u_j? (ii) What is the expected number of trials before absorption? (iii) How often will the chain be in state u_k (nonabsorbing) before absorption?

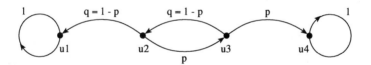

Figure 11.1: The digraph of a random walk chain.

We can answer these questions by considering

$$P^t = \begin{bmatrix} I & O \\ R & Q \end{bmatrix}^t = \begin{bmatrix} I & O \\ R_t & Q^t \end{bmatrix},$$

where $R_t = R_{t-1} + Q^{t-1}R = R + QR_{t-1}$, and $R_1 = R$.

We take $p = \frac{1}{3}$ in the example of Figure 11.1. Then, we find that

$$R = \begin{bmatrix} \frac{2}{3} & 0 \\ 0 & \frac{1}{3} \end{bmatrix}, \quad Q = \begin{bmatrix} 0 & \frac{1}{3} \\ \frac{2}{3} & 0 \end{bmatrix},$$

$$Q^2 = \begin{bmatrix} \frac{2}{9} & 0 \\ 0 & \frac{2}{9} \end{bmatrix}, \dots, Q^{2n} = \begin{bmatrix} \left(\frac{2}{9}\right)^n & 0 \\ 0 & \left(\frac{2}{9}\right)^n \end{bmatrix}, \quad Q^{2n+1} = \begin{bmatrix} 0 & \frac{1}{3}\left(\frac{2}{9}\right)^n \\ \left(\frac{2}{9}\right)^n \frac{2}{3} & 0 \end{bmatrix},$$

and, thus,

$$Q^t \rightarrow \begin{bmatrix} 0 & 0 \\ 0 & 0 \end{bmatrix} \text{ as } t \rightarrow \infty.$$

We next observe that $(I - Q)$ is invertible. Indeed, we find the inverse to be explicitly

$$(I - Q)^{-1} = \begin{bmatrix} 1 & -\frac{1}{3} \\ -\frac{2}{3} & 1 \end{bmatrix}^{-1} = \begin{bmatrix} \frac{9}{7} & \frac{3}{7} \\ \frac{6}{7} & \frac{9}{7} \end{bmatrix}.$$

This can also be expressed as a series in Q^m by using the fact that $\frac{1}{1-r} = \sum_{n=0}^{\infty} r^n$ for $|r| < 1$:

$$\sum_{m=0}^{\infty} Q^m = \sum_{n=0}^{\infty} Q^{2n} + \sum_{n=0}^{\infty} Q^{2n+1}$$

$$= \sum_{n=0}^{\infty} \begin{bmatrix} \left(\frac{2}{9}\right)^n & 0 \\ 0 & \left(\frac{2}{9}\right)^n \end{bmatrix} + \sum_{n=0}^{\infty} \begin{bmatrix} 0 & \frac{1}{3}\left(\frac{2}{9}\right)^n \\ \frac{2}{3}\left(\frac{2}{9}\right)^n & 0 \end{bmatrix}$$

$$= \begin{bmatrix} \frac{1}{1-\frac{2}{9}} & 0 \\ 0 & \frac{1}{1-\frac{2}{9}} \end{bmatrix} + \begin{bmatrix} 0 & \frac{1}{3}\frac{1}{1-\frac{2}{9}} \\ \frac{2}{3}\frac{1}{1-\frac{2}{9}} & 0 \end{bmatrix}$$

$$= \begin{bmatrix} \frac{9}{7} & \frac{3}{7} \\ \frac{6}{7} & \frac{9}{7} \end{bmatrix}.$$

Thus, in this case, we have

$$(I - Q)^{-1} = \sum_{m=0}^{\infty} Q^m, \tag{11.1}$$

which can be used to find the limit of R_t as $t \to \infty$ as well. By repeating the second equation for R_t, we find that

$$
\begin{aligned}
R_t &= R + QR + Q^2 R + \cdots + Q^{t-1} R \\
&= (I + Q + Q^2 + \cdots + Q^{t-1}) R \\
&= \left(\sum_{m=0}^{t-1} Q^m \right) R.
\end{aligned}
\tag{11.2}
$$

Since the limit of the right side exists, we find that

$$
\begin{aligned}
\lim_{t \to \infty} R_t &= \lim_{t \to \infty} \left(\sum_{m=0}^{t-1} Q^m R \right) \\
&= \left(\lim_{t \to \infty} \sum_{m=0}^{\infty} Q^m \right) R = (I - Q)^{-1} R.
\end{aligned}
\tag{11.3}
$$

Thus, we not only find the limit exists but its value as well from this formula,

$$
R_\infty := \lim_{t \to \infty} R_t = (I - Q)^{-1} R = \begin{bmatrix} \frac{9}{7} & \frac{3}{7} \\ \frac{6}{7} & \frac{9}{7} \end{bmatrix} \begin{bmatrix} \frac{2}{3} & 0 \\ 0 & \frac{1}{3} \end{bmatrix}
$$

$$
= \begin{bmatrix} \frac{6}{7} & \frac{1}{7} \\ \frac{4}{7} & \frac{3}{7} \end{bmatrix}.
$$

We now return to the original 4×4 matrix and find that

$$
\lim_{t \to \infty} P^t = \begin{bmatrix} I & O \\ R_\infty & O \end{bmatrix} = \begin{bmatrix} 1 & 0 & 0 & 0 \\ 0 & 1 & 0 & 0 \\ \frac{6}{7} & \frac{1}{7} & 0 & 0 \\ \frac{4}{7} & \frac{3}{7} & 0 & 0 \end{bmatrix}.
$$

This enables us to answer the first question posed. The probability of being absorbed in state k_1 given that the chain started in k_2 is $\frac{6}{7}$, and given that it started in k_3, the probability is $\frac{1}{7}$. Similarly, the probabilities of being absorbed in state k_4 are respectively $\frac{4}{7}$ and $\frac{3}{7}$.

This particular model is also referred to as the *gambler's ruin*, which gives a nice feeling for the model. A gambler has either 1 or 2 dollars to play with. His probability of winning a game is $\frac{1}{3}$ and of losing is $\frac{2}{3}$. He plays repeatedly until he gets 3 dollars or loses everything. If he starts with 1 dollar, the probability of losing everything is $\frac{6}{7}$, but if he started with 2 dollars, it is only $\frac{4}{7}$.

This, of course, is only a simplified version, but can be extended to any number of states corresponding to the various possible holdings of a gambler. However, it does illustrate one rule: If you gamble, it's better to be rich.

The results here refer only to this example. It is somewhat surprising that they hold in general as well.

11.2   Some General Results

THEOREM 11.1
In an absorbing Markov chain with states ordered such that the transition matrix has the form

$$P = \begin{bmatrix} I & O \\ R & Q \end{bmatrix},$$

the following hold:

(i) $Q^t \to 0$ *as* $t \to \infty$.

(ii) $(I - Q)$ *has an inverse given by* $(I - Q)^{-1} = \sum\limits_{m=0}^{\infty} Q^m$.

(iii)   *If* $R_t = R_{t-1} + Q^{t-1}R$, *then* $R_t \to R_\infty$ *as* $t \to \infty$.

(iv) $R_\infty = (I - Q)^{-1}R$.

PROOF   of (i): For each nonabsorbing state u_j, there is an absorbing state u_i with a path from u_j to u_i of minimum length. Let r be the maximum length of all such paths. Therefore, in r steps, there is a positive probability p of entering one of the absorbing states regardless of where you started. The probability of not reaching an absorbing state in r steps is $(1 - p)$. After the next r steps, it is $(1-p)^2$, and after kr steps, $(1-p)^k$. Since this approaches 0 as $k \to \infty$, the probability of being in any nonabsorbing state approaches 0 as $t \to \infty$. But the elements of Q^t are just these probabilities. ∎

PROOF   of (ii): We first show $\det(I-Q) \neq 0$ and, hence, $I-Q$ is invertible.

$$(I-Q)(I+Q+Q^2+\cdots+Q^m) = I-Q+Q-Q^2+Q^2-\cdots-Q^{m+1} = I-Q^{m+1}.$$

For m sufficiently large,

$$\det(I - Q^{m+1}) \geq 1 - \epsilon$$

since $I - Q^{m+1} \to I$ by (i) and det is continuous. Thus,

$$\det(I - Q)\det(I + Q + \cdots + Q^m) \geq 1 - \epsilon$$

and $\det(I - Q) \neq 0$. Moreover,

$$I = \lim_{m \to \infty} (I - Q^{m+1}) = (I - Q) \left(\sum_{m=0}^{\infty} Q^m \right)$$

from which (ii) follows. ∎

PROOF of (iii): We use the fact that $R_{t+1} - R_t = Q^t R$, and hence,

$$R_{t+2} - R_{t+1} = Q^{t+1} R,$$

$$\vdots$$

$$R_{t+m} - R_{t+m-1} = Q^{t+m-1} R.$$

Then, by adding all these terms together, we get

$$R_{t+m} - R_t = (Q^t + \cdots + Q^{t+m-1}) R$$
$$= (I + Q + \cdots + Q^{m-1}) Q^t R.$$

Since it can be shown that

$$\sum_{k=0}^{m-1} Q^k = (I - Q^m)(I - Q)^{-1}$$

by multiplying both sides by $(I - Q)$, it follows that $\sum_{k=0}^{m} Q^k$ is bounded uniformly in m. This implies that

$$R_{t+m} - R_t = \sum_{k=0}^{m-1} Q^k Q^t R \to 0 \text{ as } t \to \infty$$

for $m \geq 0$ and similarly for $m < 0$. Thus, $\{R_t\}$ is a Cauchy sequence and so is $\{r_{ij}(t)\}$, the sequence of its elements which therefore must converge to some real numbers $r_{ij}(\infty)$ as $t \to \infty$. ∎

Part (iv) follows from (11.3), which holds in the general case as well.

This theorem shows that the answer to the first question posed at the start of this chapter is the same in general as in the example and is given by R_∞. The second question, the expected number of trials before absorption, has been answered neither for the special case nor in general. Here is the answer.

THEOREM 11.2
The expected number of times a chain is in the nonabsorbing state k_j given that it started in k_i is given by the corresponding element of $(I - Q)^{-1}$.

Before we give the proof, we consider the meaning in a few examples. In Russian roulette, we have already discussed the result in Chapter 1. The matrix $Q = [5/6]$, $I - Q = [1/6]$ and, therefore, $(I - Q)^{-1} = [6]$; that is, the expected (average) number of times we can play the game before being absorbed (dead).

For the example in Section 11.1, we have already found $(I - Q)^{-1}$. If we begin with \$1, we can expected to have \$1 on the average for 9/7 games before we either go broke or win \$3. If we start with \$2, we can expect to have \$1 on the average for 6/7 games. The expression *expected value* means a weighted average. It is usually applied to random variable X and uses the notation $E(X)$. If X takes the value x_i with probability p_i, $i = 1, \ldots, n$, then

$$E(X) = \sum_{i=1}^{n} x_i p_i, \qquad (11.4)$$

i.e., it is just the weighted average of the values $x_i, i = 1, \ldots, n$. The simplest case is a Bernoulli random variable which takes the values 0 or 1 with probabilities q and $p = 1 - q$. Then, $E(X) = 0 \cdot q + 1 \cdot p = p$. For example, in a fair coin, with 0 corresponding to tails and 1 to heads, the expected number of heads in a toss of the coin is $E(X) = \frac{1}{2}$. In repeated tosses of a coin, we have a sequence $X_1, X_2, \ldots, X_n$ of independent random variables. They have the property that

$$E \left(\sum_{i=1}^{n} X_i \right) = \sum_{i=1}^{n} E(X_i) = n \cdot \frac{1}{2},$$

so if we toss a coin 10 times, we expect to get 5 heads.

PROOF OF THEOREM. Let $X_{ij}(t)$ be the random variable with value 1 if the chain (at time t) is in state u_j, having started in u_i, and with value 0 otherwise. The probability that $X_{ij}(t)$ is 1 is just $q_{ij}(t)$, the element corresponding to the u_i row and u_j column of Q^t. Since this is just a Bernoulli random variable, we see the expected value is

$$E(X_{ij}(t)) = q_{ij}(t),$$

which we now sum over all time to get

$$E \left(\sum_{t=0}^{\infty} X_{ij}(t) \right) = \sum_{t=0}^{\infty} E(X_{ij}(t)) = \sum_{t=0}^{\infty} q_{ij}(t).$$

This infinite sum $Y_{ij} = \sum_{t=0}^{\infty} X_{ij}(t)$ is the number of times the chain is in state u_j, having started in state u_i, which is just what we want. If we denote by y_{ij} the elements of $(I - Q)^{-1}$ and use the fact that

$$\sum_{t=0}^{\infty} Q^t = (I - Q)^{-1},$$

we conclude that

$$E(Y_{ij}) = \sum_{t=0}^{\infty} E(Z_{ij}(t)) = \sum_{t=0}^{\infty} q_{ij}(t) = y_{ij}.$$

∎

COROLLARY 11.3

The expected number of steps before absorption given that the state started in u_i is the sum of the row of $(I - Q)^{-1}$ corresponding to u_i.

Problems 11.1

1. For each of the following transition matrices of absorbing Markov chains, put into standard form and then find Q, R, and R_∞.

   a. $P = \begin{bmatrix} \frac{2}{3} & \frac{1}{3} \\ 0 & 1 \end{bmatrix}$,

   b. $P = \begin{bmatrix} 0 & 0 & 1 \\ \frac{1}{3} & \frac{1}{3} & \frac{1}{3} \\ 0 & 0 & 1 \end{bmatrix}$,

   c. $\begin{bmatrix} 0 & 0 & \frac{1}{3} & \frac{2}{3} \\ 0 & 1 & 0 & 0 \\ 0 & 0 & 1 & 0 \\ \frac{1}{2} & \frac{1}{4} & \frac{1}{4} & 0 \end{bmatrix}$.

2. Find the expected number of trials in which the chain is in each of the nonabsorbing states given that it starts in another in each of the Problems 1a, 1b, and 1c.

11.3 Population Genetics: An Example of an Absorbing Chain

In Mandelian or population genetics, a trait such as color of eyes, length of nose, and crookedness of teeth is passed from one generation to the next by means of genes. Each trait is governed by (at least) two genes, each of which can have two forms (alleles), one of which is dominant A and the other is recessive a. An individual has *genotype* AA, Aa, or aa. The *phenotype* or observed trait is always that of the dominate allele; e.g., if A corresponds to brown eyes and a to blue eyes, then both an AA and an Aa individual will have brown eyes while aa has blue eyes.

An individual receives one gene from each parent and passes on one at random to each offspring. Thus, if the parents are AA and Aa genotype, then half of the offspring will be AA and half Aa. If both the parents are Aa, then $\frac{1}{4}$ of the offspring will be AA, $\frac{1}{2}$ Aa, and $\frac{1}{4}$ aa.

Based on these rules, one can construct Markov chain models for various reproduction procedures. For example, if the procedure is to continually cross-breed with a recessive individual, the chain of Figure 11.2 results with

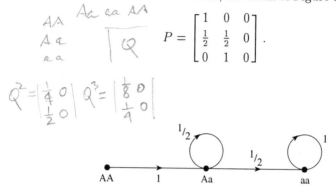

$$P = \begin{bmatrix} 1 & 0 & 0 \\ \frac{1}{2} & \frac{1}{2} & 0 \\ 0 & 1 & 0 \end{bmatrix}.$$

Figure 11.2: Cross-breeding with a recessive.

This is an absorbing chain with $P = \begin{bmatrix} I & O \\ R & Q \end{bmatrix}$, where $Q = \begin{bmatrix} \frac{1}{2} & 0 \\ 1 & 0 \end{bmatrix}$ and $(I - Q)^{-1} = \begin{bmatrix} 2 & 0 \\ 2 & 1 \end{bmatrix}$. An individual in state Aa will be absorbed by aa in two generations $(2 + 0)$, whereas one in AA will be absorbed in three generations $(2 + 1)$ on the average.

In inbreeding, two individuals of opposite sex are chosen at random and two of their offspring are chosen at random and mated. This is repeated. The states now are genotypes of pairs of individuals of which there are six (not ordered):

$$u_1 = (AA, AA), \quad u_2 = (aa, aa), \quad u_3 = (AA, Aa),$$
$$u_4 = (AA, aa), \quad u_5 = (Aa, Aa), \quad u_6 = (Aa, aa).$$

The resulting Markov chain has the digraph given by Figure 11.3. Notice that the chain now has two absorbing states.

Problems 11.3

1. Construct the Markov chain corresponding to repeated cross-breeding with a hybrid.

2. Find the matrix in the form $P = \begin{bmatrix} I & O \\ R & Q \end{bmatrix}$ for the inbreeding process given by the digraph in Figure 11.3. Find R_∞ and $(I - Q)^{-1}$. (You may wish to use Maple for this.)

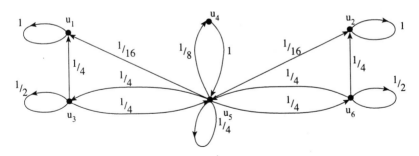

Figure 11.3: The chain resulting from inbreeding.

3. Give a genetic interpretation of the meaning of the elements in R_∞ and $(I - Q)^{-1}$. Comment on the difference between u_1 and u_5, both of which have the same phenotype. Suppose a breeder starts with a pair which have the same phenotype, but different genotypes; how can he tell?

11.4 Small Group Decision Making: An Absorbing Markov Chain

A jury of six persons must decide on a verdict unanimously. They have three choices: G_1, guilty of 1st degree, G_2, guilty of 2nd degree, and NG, not guilty. Whenever one of the jurors announced that he has changed his mind, a new vote is taken. This is continued until the vote is $(6, 0, 0)$. There are seven possible votes: $w_1 = (6, 0, 0)$, $w_2 = (5, 1, 0)$, $w_3 = (4, 2, 0)$, $w_4 = (4, 1, 1)l$, $w_5 = (3, 3, 0)$, $w_6 = (3, 2, 1)$, and $w_7 = (2, 2, 2)$, where the first number is the verdict with the largest vote, the second next largest, and so forth.

The digraph of the voting is given in Figure 11.4.

This is an absorbing chain with w_1 as the absorbing state. If a juror who changes his mind is equally likely to choose the other two alternatives, how long, if ever, will it take to reach unanimity? The digraph is under the assumption that the person who changes his mind is equally likely to choose the other two verdicts. Another assumption is that each person is equally likely to change his mind.

A more tractable problem is the following: Suppose the jury has only four persons. Then, the possible votes are

$$w_1 = (4, 0, 0), \quad w_2 = (3, 1, 0), \quad w_3 = (2, 2, 0), \quad w_4 = (2, 1, 1).$$

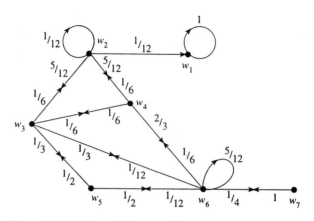

Figure 11.4: Digraph for small group decision making.

The digraph is as in Figure 11.5.

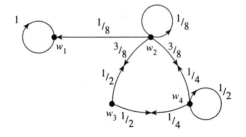

Figure 11.5: The digraph for a four person jury.

Then,

$$P = \begin{array}{c} \\ w_1 \\ w_2 \\ w_3 \\ \\ \end{array}\begin{array}{cccc} w_1 & w_2 & w_3 & w_4 \\ \end{array} \left[\begin{array}{cccc} 1 & 0 & 0 & 0 \\ \frac{1}{8} & \frac{1}{8} & \frac{3}{8} & \frac{3}{8} \\ 0 & \frac{1}{2} & 0 & \frac{1}{2} \\ 0 & \frac{1}{4} & \frac{1}{4} & \frac{1}{2} \end{array}\right]$$

if each person is equally likely to change his mind and is equally likely to choose the other two alternatives.

Problems 11.4

1. Find $(I - Q)^{-1}$ for the four-person jury, and assuming that chain starts in w_4, find expected number of mind changes to absorption.

2. Explain how the probabilities in Figure 11.5 are obtained under the assumptions given.

3. Express the transition matrix of the chain in Figure 11.4 in standard form. Then, use Maple or other program to find how many steps are needed to go from a deadlocked jury $(2, 2, 2)$ to $(6, 0, 0)$.

$$(1) \quad (1-Q) = \begin{vmatrix} 1 & -\frac{1}{2} \\ -\frac{1}{8} & -\frac{3}{8} \end{vmatrix} \quad (1-Q)^{-1} =$$

12

From Markov Chains to Compartmental Models

For our Markov chain models, we have always assumed that the state vectors are probabilities. However, in some applications such as ecosystems, we are more interested in the quantities of, say, a nutrient in a particular state at a particular time rather than the probability that a molecule is in that state at that time. We shall see that the two concepts are equivalent.

12.1 Comparison of Quantities

The probability $p_i(n)$ that the molecule is in state i at time n may be found from the levels of material $x_i(n)$ in that state (now called compartment) at time n. Indeed $p_i(n) = x_i(n) / \sum_{j=1}^{k} x_j(n)$, where k is the number of states. The levels at time $n + 1$ are given in terms of those at time n by the same equation,

$$\mathbf{x}_{n+1}^T = \mathbf{x}_n^T P, \quad n = 0, 1, \ldots, \tag{12.1}$$

as the probabilities. Here, $\mathbf{x}_n$ is a column vector of material levels. Then, we have

$$\mathbf{x}_{n+1}^T \begin{bmatrix} 1 \\ 1 \\ \vdots \\ 1 \end{bmatrix} = [x_{n+1,1}, x_{n+1,2}, \ldots, x_{n+1,k}] \begin{bmatrix} 1 \\ 1 \\ \vdots \\ 1 \end{bmatrix}$$

$$= \mathbf{x}_n^T P \begin{bmatrix} 1 \\ 1 \\ \vdots \\ 1 \end{bmatrix} = \mathbf{x}_n^T \begin{bmatrix} 1 \\ 1 \\ \vdots \\ 1 \end{bmatrix} \tag{12.2}$$

since $[1, 1, \ldots, 1]^T$ is always a right eigenvector corresponding to the steady state eigenvalue of 1 of P. If we start with a quantity $q = \sum_{j=1}^{n} x_j(0)$ of materials in the system, then the total quantity in the system remains at q for all time by (12.2). Thus, we have $p_i(n) = \frac{x_i(n)}{q}$.

Thus, Equation (12.1) is one form of the equation of a compartmental system, but a more common format is as a difference equation

$$\mathbf{x}_{n+1}^T - \mathbf{x}_n^T = \mathbf{x}_n^T(P - I)$$

or by taking transposes of the matrices

$$\Delta \mathbf{x}_n = (P^T - I)\mathbf{x}_n. \tag{12.3}$$

If the time step, i.e., the time between trials, is h rather than 1, then $\mathbf{x}_n = \mathbf{x}(nh)$ and the left side of (12.3) is replaced by the difference quotient

$$\frac{\mathbf{x}(nh + h) - \mathbf{x}(nh)}{h} = \frac{1}{h}(P^T - I)\mathbf{x}(nh) := A\mathbf{x}(nh).$$

We now let $t = nh$ to get

$$\frac{\mathbf{x}(t + h) - \mathbf{x}(t)}{h} = A\mathbf{x}(t).$$

This left side is approximately the derivative, so we have

$$\mathbf{x}' \approx A\mathbf{x}. \tag{12.4}$$

This is the differential equation for a compartmental model. Notice that the matrix $A = \frac{1}{h}(P^T - I)$ has all its column sums equal to zero, has nonpositive diagonal elements, and has non-negative off-diagonal elements. Such matrices are called compartmental matrices. The compartmental model then is the system of equations with initial condition

$$\mathbf{x}' = A\mathbf{x}, \quad \mathbf{x}(0) = \mathbf{x}_o. \tag{12.5}$$

The nondiagonal coefficients in A are the flow rates between compartments. They may be obtained directly from a weighted digraph (see Figure 12.1).

Notice that, except for the loops, the two digraphs are the same. The weights, however, are different; those for the Markov chain are probabilities, whereas those for the compartmental model are relative flow rates. In this example, the transition matrix is

$$P = \begin{bmatrix} .6 & .4 & 0 \\ .1 & .7 & .2 \\ .1 & 0 & .9 \end{bmatrix}$$

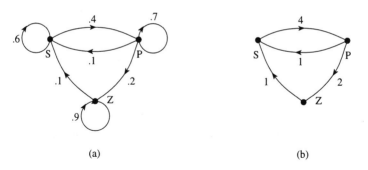

(a)   (b)

Figure 12.1: The weighted digraph of the closed aquatic ecosystem example of Problem 3 of Section 10.2 corresponding to the Markov chain model (a) and the compartmental model $(h = 0.1)$ (b).

and the compartmental matrix for $h = 0.1$ is

$$A = \begin{bmatrix} -4 & 1 & 1 \\ 4 & -3 & 0 \\ 0 & 2 & -1 \end{bmatrix}.$$

We saw that all P have an eigenvalue 1 with the right eigenvector $[1, 1, \ldots, 1]^T$. For the compartmental matrix, since the columns add to 0, 0 is always an eigenvalue, but the eigenvector corresponding to it is the left eigenvector $[1, 1, \ldots, 1]$.

The fixed-point probability vector, $\mathbf{w}$, for a singular Markov chain, as in Figure 12.1, corresponds to the equilibrium solution of the differential equation that is the solution to (12.5) for which $\mathbf{x} = 0$. This is also the asymptotic solution, as we shall see in Part IV.

Although Markov chains and discrete compartmental models of the type considered here are equivalent, there are other types of compartmental models which are not. The flows may depend on several compartments or on none at all. The functional form of the flows may be nonlinear, and in some examples, it must be. There is also the possibility of having flows from and to the outside, which is not representable by a digraph. The analysis of compartmental models is thus more general than Markov chains. However, certain results of the latter will be used with the compartmental model theory.

In compartmental models, it is traditional to present the digraph in the form of "box diagram" in which the vertices are boxes instead of points (see Figure 12.2).

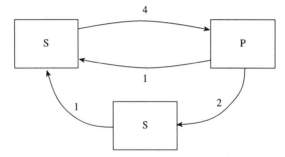

Figure 12.2: The compartmental model of Figure 12.1 presented as a "box diagram."

12.2   Two Examples

In this section, we consider a small ecosystem for which both compartmental models and Markov chain models are constructed. We shall trace both their energy and nutrient flows through the system. The compartments are the same, but the results are quite different.

This is generally true when the flow of energy of one of several nutrients is traced through an ecosystem. Typically, energy flow can be modeled by an open hierarchical system, whereas nutrient flow leads to a closed system. With Markov chains, open systems can be modeled only by introducing additional compartments and then not always. But compartmental models may include inputs and outputs as well.

Consider the following simple system consisting of producers (x_1), herbivores (x_2), decomposers (x_3), and abiotic elements (x_4). The energy flow begins with an input f_{10} to the primary producers. There is a flow from the producers to the herbivores and from the herbivores to the decomposers. The losses from each compartment due primarily to metabolism and respiration are represented by flows to the outside. This model of energy flow is shown in Figure 12.3.

Nutrient flow, on the other hand, could be given by the model shown in Figure 12.4. Some of the nutrient in the herbivore compartment is recycled directly to the abiotic portion, whereas some is cycled through the decomposers. There is also a direct interchange between the abiotic elements and the primary producers.

These represent perhaps the simplest compartmental models which can properly be termed ecosystem models. Yet, even in these simple cases, the distinction between the energy flow and nutrient flow model becomes apparent.

The associated digraph in the first case is not connected, whereas in the second case, it is strongly connected (i.e., there is a closed path which

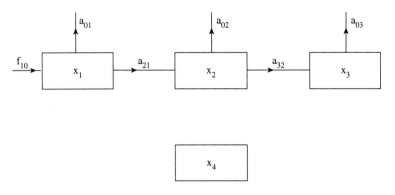

Figure 12.3: Energy flow in a simple compartmental model of an ecosystem consisting of producers (x_1), herbivores (x_2), decomposers (x_3), and the abiotic environment (x_4).

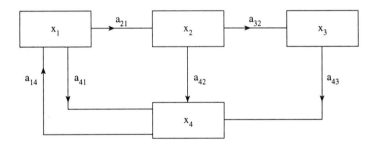

Figure 12.4: Typical nutrient flow in the ecosystem model with the same compartments as in Figure 12.1.

includes all vertices). The associated differential equations are

$$\frac{d\mathbf{x}}{dt} = \begin{bmatrix} -a_{01} - a_{21} & 0 & 0 & 0 \\ a_{21} & -a_{02} - a_{32} & 0 & 0 \\ 0 & a_{32} & -a_{03} & 0 \\ 0 & 0 & 0 & 0 \end{bmatrix} \mathbf{x} + \begin{bmatrix} f_{10} \\ 0 \\ 0 \\ 0 \end{bmatrix} \tag{12.6}$$

and

$$\frac{d\mathbf{x}}{dt} = \begin{bmatrix} -a_{21} - a_{41} & 0 & 0 & a_{14} \\ a_{21} & -a_{42} - a_{32} & 0 & 0 \\ 0 & a_{32} & -a_{43} & 0 \\ a_{41} & a_{42} & a_{43} & -a_{14} \end{bmatrix} \mathbf{x}. \tag{12.7}$$

In the first case, the eigenvalues are all negative or zero, while in the second, they all have negative real parts or are zero.

In the first case, one must ignore the input and must add another compartment for the outputs in order to construct the corresponding Markov chain. However, since it has no inputs or outputs, compartment 4 may be ignored. The transition matrix is then given by

$$P = \begin{bmatrix} 1 - h(a_{21} + a_{01}) & ha_{21} & 0 & ha_{01} \\ 0 & 1 - h(a_{02} + a_{32}) & ha_{32} & ha_{02} \\ 0 & 0 & 1 - h(a_{03}) & ha_{03} \\ 0 & 0 & 0 & 1 \end{bmatrix}, \quad (12.8)$$

where h, the time step, is sufficiently small to make the diagonal elements non-negative.

μ_2/μ_4?

μ_3/μ_4 arrow incorrect

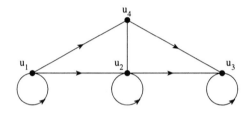

Figure 12.5: The directed graph corresponding to the Markov chain model of the energy flow of Figure 12.3.

This is the transition matrix of an absorbing Markov chain. The last row corresponds to an absorbing state and the others to transient states; that is, there is a positive probability that a unit of energy in any of the states except the last will leave it. Ultimately, all energy will be absorbed by the last state, which corresponds to the outside of the system. This, of course, is not much of a surprise and is already obvious from the digraph of Figure 12.5, where u_1, u_2, and u_3 correspond to the first three compartments and u_4 corresponds to the outside. The transition matrix (12.7) may be put in the standard form

$$P = \begin{bmatrix} I & O \\ R & Q \end{bmatrix}.$$

From the last chapter, we know that $(I - Q)$ is invertible. This is also clear in this particular case, since the inverse is given by

$$(I - Q)^{-1} = \frac{1}{h} \begin{bmatrix} \frac{1}{b_1} & \frac{a_{21}}{(b_1 b_2)} & \frac{a_{21}a_{32}}{(b_1 b_2 a_{03})} \\ 0 & \frac{1}{b_2} & \frac{a_{32}}{b_2 a_{03}} \\ 0 & 0 & \frac{1}{a_{03}} \end{bmatrix}, \quad (12.9)$$

where $b_1 = a_{01} + a_{21}$ and $b_2 = a_{02} + a_{32}$. Again, from the theory of absorbing Markov chains, the elements in the first row are the expected

To Tal because The system is Catenary

time in each compartment before absorption (in units of h steps). In the original time units, it is the same but with the quotient h omitted. The total expected time to absorption is

p. 95

$$E = \frac{1}{b_1} + \frac{a_{21}}{b_1 b_2} + \frac{a_{21} a_{32}}{b_1 b_2 a_{03}}.$$

We turn now to the corresponding compartmental model given by Figure 12.3 and the associated differential equation (12.6). In this case, since the matrix is triangular, it is simple to solve the equation. In fact, the eigenvalues are just the diagonal elements of the matrix; hence, the solution is

p. 234

$$\mathbf{x}(t) = K e^{\Lambda t} K^{-1} \mathbf{x}(0) + \int_0^t K e^{\Lambda(t-s)} K^{-1} \mathbf{f}(s) \, ds, \qquad (12.10)$$

where $\mathbf{f} = [f_{10}, 0, 0, 0]^T$, K is the matrix of eigenvectors, and $e^{\Lambda t}$ is the matrix with $e^{\lambda_i t}$'s on the diagonal and 0's elsewhere. (See Appendix.) In this case, we may omit the fourth compartment again and obtain for the first three compartments the asymptotic levels:

$$\lim_{t \to \infty} \mathbf{x}_1(t) = 0 + \lim_{t \to \infty} K_{11} \begin{bmatrix} \frac{e^{\lambda_1 t}-1}{\lambda_1} & 0 & 0 \\ 0 & \frac{e^{\lambda_2 t}-1}{\lambda_2} & 0 \\ 0 & 0 & \frac{e^{\lambda_3 t}-1}{\lambda_3} \end{bmatrix}$$

obscure

$$= 0 - K_{11} \Lambda_{11}^{-1} K_{11}^{-1} \mathbf{f}_1 = -A_{11}^{-1} \mathbf{f}_1,$$

where K_{11}, Λ_{11}, and A_{11} are partitions (3×3) of the matrices K, Λ, and A, respectively. Since $\mathbf{f}_1^T = [f_{10}, 0, 0]$, the value in (12.10) is f_{10} times the first column of $-A_{11}^{-1}$, and since $-A_{11} = (I-Q)/h$, this first column is the same as the first row of (12.9). Thus, the energy distribution with a constant rate of input is the same as the expected time in each of the compartments before absorption.

The behavior of the Markov chain and differential equation for the nutrient flow is completely different. Since the digraph is strongly connected, the Markov chain is regular (for small h). Hence, P^n converges as $n \to \infty$ to a matrix W, all of whose rows are the same. Each row is an eigenvector of P corresponding to the eigenvalue 1, as we saw in Chapter 10. Hence, since $P = I + hA^T$, $\mathbf{w}^T$ is an eigenvector of A as well corresponding to the eigenvalue 0, i.e.,

$$A\mathbf{w}^T = \begin{bmatrix} -a_{21} - a_{41} & 0 & 0 & a_{14} \\ a_{21} & -a_{42} - a_{32} & 0 & 0 \\ 0 & a_{32} & -a_{43} & 0 \\ a_{41} & a_{42} & a_{43} & -a_{14} \end{bmatrix} \begin{bmatrix} w_1 \\ w_2 \\ w_3 \\ w_4 \end{bmatrix} = 0. \quad (12.11)$$

We know from our theory that all the components of $\mathbf{w}$ must be positive. In this example, we can show it directly. From the first equation, it is seen that w_1 and w_4 have the same sign, from the second that w_1 and w_2 have the same sign, and from the third that w_2 and w_3 do. Hence, all components have the same sign, and if one is zero, all are zero.

This same vector $\mathbf{w}$ arises when the differential equation is considered. Indeed, the matrix A has exactly one zero eigenvalue and three with negative real parts. Hence, the general solution is of the form

$$\mathbf{x}(t) = Ke^{\Lambda t}K^{-1}\mathbf{x}(0)$$
$$= c_1 e^{\lambda_1 t}\mathbf{k}_1 + c_2 e^{\lambda_2 t}\mathbf{k}_2 + c_3 e^{\lambda_3 t}\mathbf{k}_3 + c_4\mathbf{k}_4. \qquad (12.12)$$

As $t \to \infty$, $\mathbf{x}(t)$ approaches $c_4\mathbf{k}_4$, which is also an equilibrium solution to Equation (12.7), since $\mathbf{k}_4$ is an eigenvector of A belonging to the eigenvalue 0. Since $\mathbf{w}^T$ is also an eigenvector, and since the multiplicity of 0 is 1, $c_4\mathbf{k}_4$ is just some multiple of $\mathbf{w}^T$, say a. To evaluate a, we sum all the components of both sides of (12.12):

$$x_1(t) + x_2(t) + x_3(t) + x_4(t) = [1, 1, 1, 1]\mathbf{x}(t)$$
$$= [1, 1, 1, 1]\sum_{i=1}^{3} c_i e^{\lambda_i t} bfk_i$$
$$+ a[1, 1, 1, 1]\mathbf{w}^T. \qquad (12.13)$$

However, each of the columns of A adds up to 0; hence, since $\lambda_i \mathbf{k}_i = A\mathbf{k}_i$, the sum of all the components of $\mathbf{k}_i$ is zero. Since $\mathbf{w}^T$ is a probability vector, its components all add up to 1 and we have the result

$$a = \sum_{i=1}^{4} x_i(t). \qquad (12.14)$$

This is not altogether too surprising, since the system is closed and always contains the same total amount of nutrient a. What is surprising is that this nutrient will be redistributed in such a way that each compartment will ultimately end up with a positive share.

Part III

Compartmental Models: Applications

We will be concerned mainly with the construction of mathematical models that are useful in ecology, and most of our applications will be drawn from that subject. However, the same models arise in a number of other subjects such as biomedicine, physiology, economics, genetics, and business, and some of our examples will be chosen from these subjects.

Compartmental models in one form or another are the ones most commonly used for analyzing an ecosystem. They are used for description, for simulation on a computer, and for theoretical analysis, depending on the complexity and the accuracy of the data. Moreover, many other traditional models, such as the logistic equation, Lotka–Volterra equations, Markov chains, and Leslie matrices can be interpreted as compartmental models. Thus, they unify a number of diverse theories.

13

Introduction to Compartmental Models

In this chapter, we first present elements of the theory of compartmental models. We then present a few special cases and examples and examine the structure of the associated matrices.

13.1 Basic Concepts

The construction of a compartmental model is straightforward and highly intuitive. The ecosystem (or any other system) is partitioned into homogeneous compartments and the flow of nutrients or energy (or of money, goods, electrons, radioactive tracers, information, etc.) between them is traced. In order to keep the discussion general, we shall refer to the *flow of material* between compartments, with the understanding that it could be any of the materials mentioned. Similarly, we shall be concerned with the *level of material* in each compartment. The compartments are represented by boxes and the flows by arrows (see, e.g., Figure 13.1).

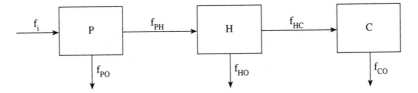

Figure 13.1: The flow of energy through a simple ecosystem. The energy flow from the sun (f_i) is taken up by plants (P), some of which are eaten by herbivores (H), which, in turn, are eaten by carnivores (C). Part of the energy flows to the outside of the system (e.g., f_{PO}).

This representation is the graphical or structural aspect of the compartmental model. Some information can be gleaned from it at this stage by considering it to be a directed graph in which the compartments are vertices and the flows are arcs. The outside must also be considered as an additional vertex for this to give a proper digraph.

However, in order to simulate or analyze a system, the model must be quantified. This is done by studying the rate of change of the level x_i of the ith compartment in time. This quantity will vary over time and will be written as

$$x_i = x_i(t) \tag{13.1}$$

when the time dependence must be made explicit. The flow between the ith and j^{th} compartment, designated f_{ij}, is a rate, measured in quantity of material per unit time.

A differential equation describing the behavior of x_i may be obtained by equating the time rate of change of x_i to the difference between the flows coming in and those going out of the ith compartment. In mathematical terms, this is

$$\begin{matrix} \dfrac{dx_i}{dt} & = \displaystyle\sum_k f_{ki} & -\displaystyle\sum_j f_{ij}. \\ \text{rate of change} & = \text{flows in} & -\text{flows out} \end{matrix} \tag{13.2}$$

The f_{ij} may be constant or variable in time and usually are functionally dependent on the x_i's.

The most widely used assumption, particularly in physiology and medicine, is that the functional form of f_{ij}, the flow rate, is

$$f_{ij} = a_{ij} x_j.$$

This corresponds to a certain fraction a_{ij} of the level of material x_i in compartment i passing to compartment j per unit time. In the ecological literature, this is often called *donor controlled flow*. The a_{ij} may be functions of time but are usually assumed to be independent of the x_i's.

Other common assumptions for the functional form of the flow rates are

$$\begin{aligned} f_{ij}(1) &= c_{ij} & &\text{(constant)}, \\ f_{ij}(2) &= a_{ij} x_i & &\text{(donor control)}, \\ f_{ij}(3) &= b_{ij} x_j & &\text{(recipient control)}, \\ f_{ij}(4) &= d_{ij} x_i x_j & &\text{(Lotka–Volterra)}, \\ f_{ij}(5) &= \alpha_{ij} x_i/(\beta_{ij} + x_i) & &\text{(chemostat)}. \end{aligned}$$

The constant flow rate is not very realistic but may be used as a rough approximation when little is known about the system or when the system is close to equilibrium. The donor and recipient control correspond, in the case of an ecosystem, to conditions of scarcity and abundance of resources, respectively. The two remaining expressions are the most common nonlinear forms. There are many other possibilities; sometimes, the flow depends on the level in other than the donor or recipient compartments and may have a more complex functional form. Also, there might be delays in the uptake of a nutrient, for example.

The equations involving nonlinear flows are often analyzed by linearization, which leads to the first three forms. Most of our analysis will be based on them.

Very often, particularly in ecosystems, the use of the derivative is inappropriate and a finite difference should be used instead. This happens, for example, if diurnal data are used. Then, Equation (13.2) is replaced by

$$\frac{\Delta x_i}{h} = \sum_k \bar{f}_{ki} - \sum_j \bar{f}_{ij}, \tag{13.3}$$

where h is the time step and $\bar{f}_{ij}$ are the flow rates averaged over the time h.

Steps in construction of a compartmental model:

1. Split up system into a number of homogeneous subsets or compartments.

2. Join compartments by arrows corresponding to flows of material.

3. Use directed graph for preliminary analysis.

4. Determine appropriate functional form of flow rates.

5. Use data to establish initial values of levels and values of flow rate functions.

6. Analyze by using Markov chain theory, differential equations, or computer simulation.

Example 13.1

In the leaky tank problem, two tanks are arranged in tandem, with the output of one tank flowing into the other. Suppose each tank is initially filled with 10 gal of water when a flow of 2 gal/min of a 1-lb./g solution of salt begins flowing into the first tank. A flow of 2 g/m of the mixture from the first tank flows into the second, and a flow of 2 g/m flows out of the second tank. What is the amount of salt in each of the two tanks at time t? Does the salt concentration reach an equilibrium level in each tank?

This is a two-compartment system with donor-controlled flow (see Figure 13.2). Let x_1 and x_2 be the amount of salt respectively in the two tanks. The two associated differential equations are linear constant coefficient first-order differential equations. The first is of the form

$$x_1' + px_1 = q,$$

which has as a solution $x_1(t) = x_1(0)e^{-pt} + q/p$; the second uses the solution of the first, but is not constant coefficient.

Suppose the output of the second tank flows back into the first and the input to the first tank is discontinued. Now what happens? The differential equations of the new model now each involve both of the variables x_1 and x_2. They may be solved by a change of variables; let y_1 be their sum and y_2 be their difference. This sum and difference are first-order equations that may be solved as above.

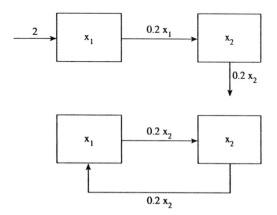

Figure 13.2: The compartmental model of the leaky tanks.

Problem 13.1

Solve the differential equations for both of the leaky tank problems using the values given. Is there an equilibrium solution in both cases?

13.2   One-Compartment Applications

A compartmental model with a single compartment seems on the face of it to be no model at all. Yet in some applications, it is useful to think of the phenomenon as a compartment with flows into and out of it. In addition, since the resulting differential equations are simple and can be solved in closed form, such a model can help to develop our intuition for the quantitative behavior of more complex models. In this section, we first consider the general behavior of linear models and then consider applications of these and nonlinear models to population dynamics and drug kinetics.

13.2.1 Linear Case

A linear one-compartmental model is as described by Figure 13.3, in which the flows f_{01} and f_{10} are linear functions of the level x in the compartment. This has a straightforward application to birth and death processes of a population of size x. The differential equation is

$$dx/dt = bx - dx, \qquad (13.4)$$

which has the solution

$$x(t) = x(0)e^{(b-d)t}.$$

Figure 13.3: A one-compartment model.

Thus, the population is exponentially increasing or decreasing depending on whether b or d is larger. In particular, if b is zero, the population will decline and approach the stable equilibrium point zero (for d not zero). If d is zero and b is not, the solution is not stable. These two cases correspond to donor and recipient control in the general case, which we cover later.

The flows may also have a constant term, in which case the equation becomes

$$dx/dt = a + bx - dx.$$

In this case, the solution is

$$x(t) = (a/(d-b))(1 - e^{(b-d)t}) + x(0)e^{(b-d)t},$$

which leads to a positive stable equilibrium solution if a and $(d-b)$ are positive.

Even more generally, we may suppose that a, b, and d are variable. The equations may still be solved, but the solutions do not have the simple form given above.

Problem 13.2

Suppose a population has an increasing birth rate, $b = c + et$, but that the death rate is constant. Find the general solution to Equation (13.4). Under what conditions does it have a stable equilibrium solution? (Divide both sides of the equation by x and then integrate them.)

13.2.2   Logistic Growth

It's clear that no population can maintain exponential growth very long since it must eventually run out of food. Rather, we would expect that both the death rate and the birth rate depend on the population size. The simplest assumption is that both are linear functions of x,

$$b = b_0 - b_1 x, \quad d = d_0 + d_1 x.$$

This is plausible since we could expect the birth rate to decrease with greater population density while the death rate increases. The equation can then be put into the form

$$dx/dt = (b_0 - d_0)x - (b_1 + d_1)x^2 = rx(1 - x/x_\infty), \qquad (13.5)$$

which is the standard *logistic equation* of population dynamics. Here, r is the *intrinsic growth rate* and x_∞ is the *carrying capacity* of the environment.

When the population is very small, the growth is approximately exponential since the term x/x_∞ is small compared to 1 and may be ignored. As the population grows and approaches x_∞, the factor $(1 - x/x_\infty)$ becomes small and the growth rate slows down. The result is the curve shown in Figure 13.4.

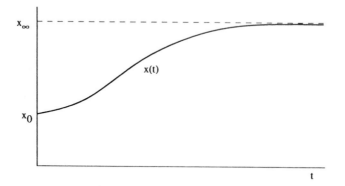

Figure 13.4: Logistic growth curve satisfying Equation (13.5).

It is also possible to solve (13.5) analytically by separating variables,

$$dx/(x(1 - x/x_\infty)) = r\,dt,$$

and then expressing the left side as

$$dx/x - dx/(x - x_\infty).$$

Both sides are then integrated to obtain

$$\ln(x) - \ln(x_\infty - x) = rt + c$$

or

$$x/(x_\infty - x) = e^c e^{rt}.$$

This expression is then solved algebraically for x:

$$x(t) = x_\infty/(1 + e^{-rt-c}). \tag{13.6}$$

Thus, in the only realistic cases, when r and x_∞ are positive, the solution converges to x_∞ as $t \to \infty$ regardless of the value of c.

This logistic equation is the basis for one model that is used in the regulation of fisheries, the so-called *Schaefer model*, which we shall cover in a subsequent chapter.

Problems 13.2

1. A cylindrical tank of radius 1 and height 1 has water flowing into its top at the rate of 2 gal/min. The water flows out at the bottom at a rate proportional to the height of water in the tank. After 1 hr, the tank is half full, having been empty at the start. How long will it be before the tank is full?

2. Same as Problem 1 except the tank is conical with radius 1 and height 1, the axis of rotation is vertical, and the tip is at the bottom.

3. A generalization of the logistic equation is the equation

$$dx/dt = rx(1 - (c+1)x/(b+cx)).$$

Find an implicit solution of this equation and use it to describe the behavior of x as t increases to infinity. The technique used to solve this equation is the same as for the logistic equation; separate the variables x and t and try to integrate each separately.

4. The population of the United States was 105 million in 1920, 150 million in 1950, and 225 million in 1980. Try to fit a logistic equation to these values by using the general formula (13.4) and solving for the three parameters r, c, and x_∞. You will find that it doesn't work. Try to explain why.

5. In tracer kinetics, the flow of material through the various compartments of the body is studied. In some cases, a one-compartment model is appropriate. For example, if a drug is injected into the blood and is eliminated directly without affecting other compartments, and then samples of blood are drawn periodically, such a model could be used. If 1 ml of the drug is injected in the blood of a person with 2

L of blood, and if it is excreted at a rate proportional to the concentration, find the amount remaining after 2 hr if after $\frac{1}{2}$ hr half of the drug remains.

13.3   Other Examples

We present here a couple of additional examples which help give a feel for the subject.

Example 13.3.1. Simple Tracer Kinetics

Tracer kinetics refers to a standard technique for studying the flow of a drug or other material through a biological system (usually a human body). The flow of the material to be studied is assumed to be near steady state. A radioactive isotope or other recognizable form of the material (the tracer) is introduced in a comparatively small quantity, so that even if the original flows are nonlinear, the tracer flows may be modeled by linear equations. These tracer experiments are among the largest consumers of compartmental models.

A simple example would be of a lipoprotein metabolism. One compartment corresponds to the blood plasma and the other to the extravascular space. A unit dose of tracer is injected into the blood and the flow follows the diagram in Figure 13.5. The quantity of interest is the flow out of compartment 1 (excretion), which, however, cannot be measured directly. Rather, only the level of compartment 1 over time is known.

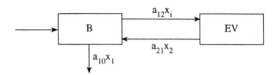

Figure 13.5: A model of lipoprotein metabolism.

The differential equations are the same as in the leaky tank model except that now there is a flow to the outside and we know the solution for $x_1(t)$. The differential equations are

$$dx_1/dt = -a_{10}x_1 - a_{12}x_1 + a_{21}x_2,$$
$$dx_2/dt = a_{12}x_1 - a_{21}x_2,$$

with initial conditions

$$x_1(0) = 1, \quad x_2(0) = 0.$$

We shall see in the next part that such equations have solutions monotonically decreasing to zero as t increases. Hence, if y denotes their sum, then y also is decreasing to zero and

$$dy/dt = -a_{10}x_1.$$

By integrating from 0 to T, we find that

$$y(T) - y(0) = \int_0^T -a_{10}x_1,$$

and since $y(0) = 1$, we have

$$a_{10} = 1/\int_0^\infty x_1.$$

Example 13.3.2. Beer Bottle Flow

Another simple example is one in which the material is composed of beer bottles. They flow between various compartments and are either all destroyed or else most are reused. Only the structure of these models is shown. The reader is invited to try to reach qualitative conclusions about the relation between them.

The compartments are (1) Bottle factory, (2) Brewery, (3) Liquor store, and (4) Drinker and the flows are as shown in Figure 13.6.

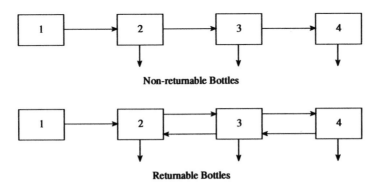

Figure 13.6: Structure and flows of a compartmental model of beer bottle flow for returnable and nonreturnable bottles.

13.4   Matrix Forms

The various functional forms of the flow rates in (13.2) can be put into matrix form, which, in some cases, simplifies the calculations. In particular, the linear cases of donor and recipient control lead to simple matrix forms. In the former, the general differential equation will be

$$d\mathbf{x}/dt = A\mathbf{x} + \mathbf{f}_i - E_o\mathbf{x} \tag{13.7}$$

where A is the matrix given by

$$A = \begin{bmatrix} a_{11} & a_{21} & \cdots & a_{n1} \\ a_{12} & a_{22} & \cdots & a_{n2} \\ \cdots & \cdots & & \cdots \\ a_{1n} & a_{2n} & \cdots & a_{nn} \end{bmatrix}$$

and E_o is the diagonal output matrix

$$E_o = \begin{bmatrix} a_{10} & 0 & \cdots & 0 \\ 0 & a_{20} & \cdots & 0 \\ \cdots & \cdots & & \cdots \\ 0 & 0 & \cdots & a_{n0} \end{bmatrix};$$

while $\mathbf{f}_i$ is the input vector. (Note that this differs from the conventional numbering of elements of A; a_{ij} is the element in the jth row and i^{th} column.)

The diagonal elements of A are given by

$$a_{ii} = -\sum_{j \neq i} a_{ij},$$

hence, each column of A must add up to zero. The matrix E_o has the proportion of flows leaving each compartment to the outside on its main diagonal, and zeros elsewhere.

In the case of recipient-controlled flow, the differential equation is similar:

$$d\mathbf{x}/dt = B\mathbf{x} + E_i\mathbf{x} - \mathbf{f}_o \tag{13.8}$$

where B has the form

$$B = \begin{bmatrix} b_{11} & -b_{21} & \cdots & -b_{n1} \\ -b_{12} & b_{22} & \cdots & -b_{n2} \\ \cdots & \cdots & & \cdots \\ -b_{1n} & -b_{2n} & \cdots & b_{nn} \end{bmatrix},$$

the b_{ii} are given by

$$b_{ii} = \sum_{j \neq i} b_{ij},$$

and the input matrix is now a diagonal matrix. In the case of matrix B as well, the columns add up to zero; hence, both A and B are singular and have a zero eigenvalue. However, the other eigenvalues behave quite differently in the two cases. As we shall see in the next section, the nonzero eigenvalues for a matrix such as A all have a negative real part, whereas for B, it is just the opposite—they have positive real part.

Most of the behavior of these two cases is already seen in the two-compartment case, which we take up next.

13.5 Two-Compartment Models

The general form of a two-compartment model with arbitrary inputs and outputs is given in Figure 13.7.

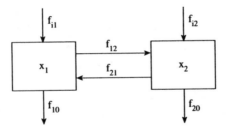

Figure 13.7: The general two-compartment model.

In this case, Equations (13.2) become, in matrix form,

$$\frac{d}{dt}\begin{bmatrix} x_1 \\ x_2 \end{bmatrix} = \begin{bmatrix} 0 & f_{21} - f_{12} \\ f_{12} - f_{21} & 0 \end{bmatrix}\begin{bmatrix} 1 \\ 1 \end{bmatrix} + \begin{bmatrix} f_{i1} \\ f_{i2} \end{bmatrix} - \begin{bmatrix} f_{1o} \\ f_{2o} \end{bmatrix}. \tag{13.9}$$

If the inputs and outputs are ignored for the moment and the flows are assumed to be donor controlled, i.e., are proportional to the levels in the donor compartments, then (13.9) becomes

$$\frac{d}{dt}\begin{bmatrix} x_1 \\ x_2 \end{bmatrix} = \begin{bmatrix} -a_{12} & a_{21} \\ a_{12} & -a_{21} \end{bmatrix}\begin{bmatrix} x_1 \\ x_2 \end{bmatrix}. \tag{13.10}$$

The recipient-controlled flows lead to an equation of the same form as Equation (13.10) except the signs of all the elements of the matrix are reversed. Hence, both donor and recipient control lead to an equation of the form

$$\frac{dx}{dt} = \begin{bmatrix} -a & b \\ a & -b \end{bmatrix} x = \mathbf{Ax}, \tag{13.11}$$

where a and b are positive in the donor-controlled case and negative in the recipient-controlled case. The eigenvalues and eigenvectors, respectively, are

$$\lambda_1 = 0, \quad \mathbf{k}_1 = \begin{bmatrix} b \\ a \end{bmatrix}; \quad \lambda_2 = -(a+b), \quad \mathbf{k}_2 = \begin{bmatrix} 1 \\ -1 \end{bmatrix}.$$

Hence, the general solution to the system may be given by

$$\mathbf{x}(t) = c_1 \begin{bmatrix} b \\ a \end{bmatrix} + c_2 \begin{bmatrix} 1 \\ -1 \end{bmatrix} e^{\lambda_2 t},$$

which may be written as

$$\mathbf{x}(t) = \begin{bmatrix} b & 1 \\ a & -1 \end{bmatrix} \begin{bmatrix} 1 & 0 \\ 0 & e^{\lambda_2 t} \end{bmatrix} \begin{bmatrix} c_1 \\ c_2 \end{bmatrix}. \tag{13.12}$$

For $t = 0$, this equation becomes

$$\mathbf{x}(0) = \begin{bmatrix} b & 1 \\ a & -1 \end{bmatrix} \begin{bmatrix} 1 & 0 \\ 0 & 1 \end{bmatrix} \begin{bmatrix} c_1 \\ c_2 \end{bmatrix},$$

which may be solved for c_1 and c_2 by inverting the matrix to get

$$\begin{bmatrix} c_1 \\ c_2 \end{bmatrix} = (-1/\lambda_2) \begin{bmatrix} 1 & 1 \\ a & -b \end{bmatrix} \mathbf{x}(0).$$

This, in turn, is substituted back into (13.12) and the three matrices are multiplied together to get

$$\mathbf{x}(t) = \begin{bmatrix} b + ae^{\lambda_2 t} & b - be^{\lambda_2 t} \\ a - ae^{\lambda_2 t} & a + be^{\lambda_2 t} \end{bmatrix} \mathbf{x}(0)/(-\lambda_2).$$

From this result, a number of conclusions follow easily:

If the system is donor controlled, then

(i) $\lambda_2 < 0$ and

$$\mathbf{x}(t) \rightarrow \begin{bmatrix} b & b \\ a & a \end{bmatrix} \mathbf{x}(0)/(-\lambda_2) = c \begin{bmatrix} b \\ a \end{bmatrix} = \mathbf{x}(\infty)$$

and this convergence is monotonic.

(ii) $\mathbf{x}(\infty)$ is a stable equilibrium solution of the differential equation.

(iii) If $\mathbf{x}(0)$ is positive, then $\mathbf{x}(t)$ is positive for all $t > 0$.

If the system is recipient controlled, then $\lambda_2 > 0$, and one or the other compartment will be annihilated in a finite amount of time. Again, $\mathbf{x}(\infty)$ is an equilibrium solution, but is now unstable. These results will be shown to hold in general in Part IV.

Problems 13.5.

1. Solve the problems of the leaky tanks in Section 13.1 by the methods of this section.

2. Suppose the model given in Figure 13.7 has donor-controlled flows between compartments and one flow to the outside, say $f_{10} = dx_1$. Find the general form of the solution for $a = b = 1$.

3. Suppose the flow from compartment 1 to compartment 2 is donor controlled, but that from 2 to 1 is recipient controlled and there are no other flows. Find the solution and the limit as $t \to \infty$.

14

Models for the Spread of Epidemics

In order to study the spread of epidemics, the population at risk is divided into compartments consisting of the number of persons with the disease (I, for infectives), those recovered from the disease and no longer susceptible (R), and those susceptible to infection (S). The population (N) is assumed to be relatively constant and that therefore $N = S + I + R$. The particular model used depends on the disease, but most can be analyzed using compartmental models.

14.1 The SIR Model

In the case of infectious diseases which are not generally fatal and which upon recovery confer immunity, such as measles, the compartmental model shown in Figure 14.1 is appropriate.

Figure 14.1: The compartmental model for a SIR epidemic.

The flow from S to I seems to be best modeled by a function given by a portion of the product of the two levels. The reason for this is that the disease spreads by the interaction of an infectious and a susceptible individual. The number of such interactions is the product. The recovery rate, however, should be donor controlled, since a portion of the infectives recover in any time period. Thus, the equations then become

$$dS/dt = -\beta IS,$$

$$dI/dt = \beta IS - \nu I,$$

$$dR/dt = \nu I.$$

Since $I+S+R = N$, the last equation is superfluous and we need consider only the first two. They have a common nonzero equilibrium point only at $(S, I) = (\nu/\beta, 0)$, which is not very interesting, since it corresponds to the no disease case. These are nonlinear equations and, hence, cannot be solved by the usual matrix approach. Rather, we try directly to find the trajectory of the solutions in the (S, I) plane in terms of S and I. The differential equation obtained by using the chain rule is

$$\frac{dI}{dS} = \frac{\frac{dI}{dt}}{\frac{dS}{dt}} = \frac{-(\beta IS - \nu I)}{\beta IS} = \left(\frac{nu}{\beta S}\right) - 1,$$

which may be solved by separating the variables to get

$$I = (\nu/\beta)\ln S - S + \text{ constant}.$$

The constant may be evaluated from the initial conditions

$$S(0) = S_0, \quad I(0) = I_0, \quad R(0) = 0$$

to get the solution

$$I = (\nu/\beta)\ln(S/S_0) - S + N \tag{14.1}$$

since $I_0 + S_0 + 0 = N$. The trajectory of these solutions looks like those in Figure 14.2, since S is always decreasing, while I is increasing for $S > \nu/\beta$ and decreasing for $S < \nu/\beta$.

Both $S(t)$ and $I(t)$ are monotone functions of t for t sufficiently large and, hence, approach values S_∞ and I_∞ as $t \to \infty$. If $S_\infty = 0$, the epidemic will have infected the entire population; if $S_\infty > 0$, the epidemic will have run its course with some of the population uninfected. The asymptotic values satisfy, by (14.1),

$$I_\infty = (\nu/\beta)\ln(S_\infty/S_0) - S_\infty + N. \tag{14.2}$$

We first show that $I_\infty = 0$, i.e., that the epidemic will eventually run its course. As the pool of susceptible individuals decreases, eventually $S < \nu/\beta - \epsilon$; then,

$$(1/I)(dI/dt) = \beta S - \nu < -\beta\epsilon.$$

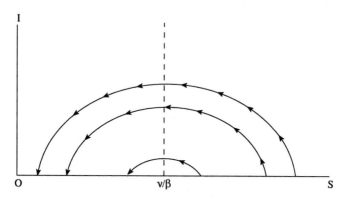

Figure 14.2: Possible trajectories of the solutions.

Both sides of this inequality may be integrated to obtain

$$\ln(I) < -\beta\epsilon t + \text{constant}$$

or, since the exponential function is increasing,

$$I(t) < C e^{-\beta\epsilon t}.$$

Hence, $I(t) \to 0$ as $t \to \infty$, i.e., $I_\infty = 0$.

The equation to be satisfied by S_∞ is obtained by expressing (14.2) as

$$S_\infty = S_0\, e^{-(\beta/\nu)(N - S_\infty)},$$

since $I_\infty = 0$. Thus, S_∞ is a zero of the function

$$f(x) = \alpha e^{\beta x/\nu} - x,$$

where $\alpha = S_0 e^{-(\beta/\nu)N}$. We note that $f(0) > 0$ and $f(N) = S_0 - N < 0$, so that $f(x)$ has a positive zero which is in the appropriate range and is unique. Thus, $S_\infty > 0$ and can never be zero. The conclusion is that under this model, the epidemic will always spare a portion of the population.

These same calculations show us that the epidemic can be avoided completely if the number of initial infectives is sufficiently small. Indeed, if $S_0 < \nu/\beta$, then $I(t)$ will be monotonically decreasing and the epidemic will never occur.

Problem 14.1

Carry out the calculations for the SIR model when $\nu = 0.01$, $\beta = 0.001$, $I_o = 1, and N = 100$.

14.2   Other Models of Epidemics

An even simpler model is the SIS model, in which the infection does not
confer immunity but rather recovered individuals again become susceptible.
The compartmental model looks like the digram in Figure 14.3.

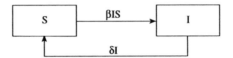

Figure 14.3: The compartmental model for an SIS epidemic.

The differential equations now reduce to a single equation since $I + S = N$,

$$dI/dt = \beta I(N - I) - \delta I = r(1 - I/K)I, \quad r = \beta N - \delta, \quad K = r/\beta.$$

This is the familiar logistic equation with intrinsic growth rate r and carrying capacity K. Its solution is

$$I = K/(1 + C e^{-rt}),$$

where C is an arbitrary constant. If we begin with one infected individual,
then $C = K - 1$ and $I(t)$ increases monotonically toward its asymptotic
value $K = r/\beta = (\beta N - \delta)/\beta = N - \delta/\beta$. Hence, the epidemic again will
run its course without infecting the entire population.

Again, there is a threshold phenomenon. If $N < \delta/\beta$, then the carrying
capacity K is negative and $I(t)$ converges to 0 monotonically.

The SIRS model has the form given in Figure 14.4. It assumes that

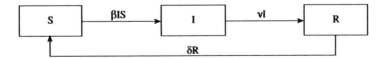

Figure 14.4: The model for an SIRS epidemic.

some of the recovered population loses its immunity and again becomes
susceptible. Again, we assume that $S + I + R = N$ is constant; hence, we

can ignore the last equation. The others are

$$dS/dt = -\beta IS + \gamma(N - I - S),$$
$$dI/dt = \beta IS - \nu I.$$

These equations are in equilibrium when $S = \nu/\beta$ and $I = \gamma(N - S)/(\gamma + \beta S)$. They may be linearized about the common equilibrium point $(\nu/\beta, \gamma(N - \nu/\beta)/(\gamma + \nu)) = (S_e, I_e)$ and the resulting linear equations may then be found and solved. This is done by letting $x = S - S_e$ and $y = I - I_e$, and then ignoring all terms except the linear terms in x and y. The two equations become

$$dx/dt = -(\beta I_e + \gamma)x - (\beta S_e + \gamma)y,$$
$$dy/dt = \beta I_e x + (\beta S_e - \nu)y. \tag{14.3}$$

which may then be solved by linear methods. (See problems.) A qualitative approach may also be used to analyze the trajectories. This is done for the similar predator-prey equations in the next chapter. Unfortunately, the equation in S and I does not have a simple closed-form solution as in the SIR model.

Problems 14.2

1. Suppose that the population is infinite. Construct the compartmental model for the SIS case and solve the equations. Explain what happens.

2. Suppose that a certain proportion of the infected population dies in the SIS case. Write down the differential equations and describe the behavior as much as possible.

3. In the SIRS model, carry out the calculations of the solutions of the linearized equations (14.3) when $N = 100$, $I_o = 10$, $\beta = 0.001$, and $\gamma = \nu = 0.01$.

15

Three Traditional Examples as Compartmental Models

The three models considered in this chapter are usually not treated as compartmental models. However, since they involve flows between compartments, they can be treated as such.

15.1 Predator–Prey or Host–Parasite Equations

For the model which describes the behavior of the population of a predator and that of its prey, we use a compartmental model with two compartments, which we denote as H and P, respectively (for host and parasite). The same symbols H and P will also stand for the respective biomasses. There is a flow of nutrients from the predator to the prey and flows into the H compartment and out of both compartments as shown in the diagram in Figure 15.1.

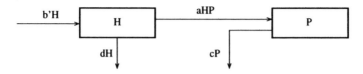

Figure 15.1: Diagram of the H-P model.

The flow into the H compartment arises from births and individual growth and, hence, is recipient controlled. The flow to the outside is the result of nonpredator deaths and excretions; the same is true for the outside flow from the predator compartment. Since the flow from H to P

depends on the interactions between the two populations, we assume it has the form aHP, which is the simplest form possible. The associated differential equations are

$$dH/dt = b'H - dH - aPH = (b - aP)H \quad (b = b' - d),$$
$$dP/dt = aHP - cP = (aH - c)P.$$

These equations, in contrast to the previous simple examples in Chapter 13, but similar to the epidemic models, are nonlinear and we cannot necessarily find a closed-form solution. However, we can study properties of the solutions; we begin with the equilibrium solutions which satisfy

$$(b - aP)H = 0, \qquad (aH - c)P = 0$$

or

$$P = b/a \text{ or } 0, \qquad H = c/a \text{ or } 0.$$

Graphically, the equilibrium solutions look like those in the P-H plane (Figure 15.2).

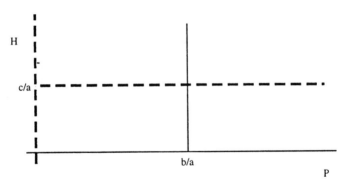

Figure 15.2: Points of equilibrium for H and P.

The common points of equilibrium are at $(P, H) = (b/a, c/a)$ or at $(0, 0)$. The two equilibrium solutions divide the P-H plane into four regions. The behavior of P and H in each of these regions is shown in the diagram of Figure 15.3. In each case, the arrows correspond to directions of increase of P or H.

Trajectories of the solutions of H and P may be of the form shown in Figure 15.4.

In order to determine what the true trajectories are, we first consider the behavior close to the common equilibrium point $(b/a, c/a)$ and denote

$$x = P - b/a, \quad y = H - c/a,$$

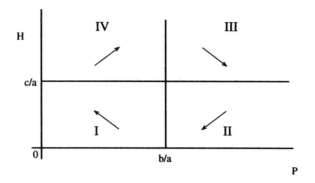

Figure 15.3: Trajectory directions.

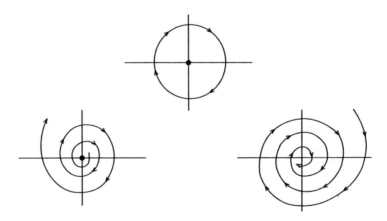

Figure 15.4: Possible trajectories of solutions.

where x and y are both small compared to H and P and their powers are even smaller. Then,

$$dy/dt = dH/dt = (b - aP_e - ax)(H_e + y) \approx -cx,$$
$$dx/dt = dP/dt = (aH_e + ay - c)(P_e + x) \approx by,$$

which are not only linear but whose form enables us to find a closed form for their trajectory. Indeed, by the chain rule, we have

$$dy/dx = (dy/dt)/(dx/dt) = -cx/by,$$

a separable equation whose solution is

$$by^2 + cx^2 = k, \quad \text{a constant.}$$

These are ellipses in the P-H plane and give a closed trajectory.

We can also find the solutions for x and y as a function of t; the equations in matrix form are

$$\frac{d}{dt}\begin{bmatrix} x \\ y \end{bmatrix} = \begin{bmatrix} 0 & b \\ -c & 0 \end{bmatrix}\begin{bmatrix} x \\ y \end{bmatrix},$$

whose eigenvalues are purely imaginary:

$$\lambda = \pm\sqrt{cb}\,i = \pm\omega i.$$

The corresponding eigenvectors are

$$K_1 = \begin{bmatrix} \sqrt{c} \\ i\sqrt{b} \end{bmatrix}, \qquad K_2 = \bar{K}_1,$$

and the solution is therefore of the form

$$x = c_1 K_1 e^{i\omega t} + c_2 K_2 e^{-i\omega t}.$$

In particular, if $C_1 = -iC_2$, then

$$x = \left(1/\sqrt{c}\right)\sin\omega t, \quad y = \left(1/\sqrt{b}\right)\cos\omega t.$$

This is consistent with our previous trajectory.

We now return to P and H and observe that the approximate solutions are

$$P \approx b/a + \left(1/\sqrt{c}\right)\sin\omega t$$
$$H \approx c/a + \left(1/\sqrt{b}\right)\cos\omega t.$$

We do not know, however, if the trajectories are closed when P and H are not close to the equilibrium point, but we can use the same trick as before to deduce that

$$dH/dP = (b - aP)H/(aH - c)P,$$

which, again, is separable. The solution is given implicitly by

$$aH - c\log H = b\log P - aP + \text{ constant},$$

which, again, is closed curve, but not as simple as an ellipse.

Problem 15.1

Imitate the procedure in this section when the compartmental model is as in Figure 15.5.

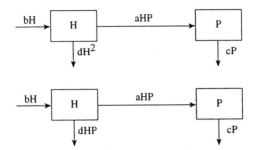

Figure 15.5: Two alternative models for predator-prey equations.

15.2   The Leslie Matrix

The Leslie matrix is used to model the age classes of a biological population. We want to find (i) the age distribution and size of the population given an initial age distribution, and (ii) conditions for stability. The population is first divided into compartments composed of age classes. The jth compartment is the number of individuals of age $j-1$ to j, $j = 1, 2, \ldots, m$. (See Figure 15.6.)

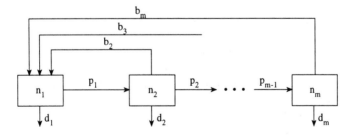

Figure 15.6: The diagram for the population model.

Time is taken to be discrete; p_i = probability of survival from age i to $i + 1$; $d_i = 1 - p_i$ = probability of dying during this same age; b_j = mean number of births to individuals in age class $j > 1$.

Let $\mathbf{N}(t) = \begin{bmatrix} n_1(t) \\ n_2(t) \\ \vdots \\ n_m(t) \end{bmatrix}$; then, $\mathbf{N}(t+1) = B\,\mathbf{N}(t)$, where B is the matrix

$$B = \begin{bmatrix} 0 & b_2 & b_2 & \cdots & b_{m-1} & b_m \\ p_1 & 0 & 0 & \cdots & 0 & 0 \\ 0 & p_2 & 0 & \cdots & 0 & 0 \\ \vdots & & & \cdots & & \vdots \\ 0 & 0 & 0 & & p_{m-1} & 0 \end{bmatrix}.$$

This is not the transpose of the matrix of a Markov chain since the columns do not add to 1. If the oldest age class has a positive mean number of births, then the matrix is nonsingular and can be inverted. This inverse can be used to find the prior age distribution given the current one.

In order to find the stable age distribution, we look for a positive eigenvalue with an associated non-negative eigenvector. This is the only kind that makes biological sense. Suppose then that λ_0 and $\mathbf{N}_0$ are such a pair. Then, if $\mathbf{N}(t) = \mathbf{N}_0$, we see that

$$\mathbf{N}(t+1) = \mathbf{B}N_0 = \lambda_0\mathbf{N}_0,$$

and if we start with the eigenvector $\mathbf{N}(0) = \mathbf{N}_0$, we get

$$\mathbf{N}(t) = \lambda_0^t\mathbf{N}_0.$$

THEOREM 15.1
If b_m is positive, then there is a unique positive eigenvalue and a stable age distribution.

PROOF   The characteristic polynomial is

$$p(\lambda) = \det|B - \lambda I| = \lambda^m - p_1 b_2 \lambda^{m-2} - p_1 p_2 b_3 \lambda^{m-3} - \cdots - p_1 p_2 \cdots p_{m-1} b_m.$$

By Descarte's rule of signs, this polynomial has exactly one positive zero, i.e., one positive eigenvalue λ_0.

By means of elementary row operations, the matrix $B - \lambda_0 I$ can be reduced to the form

$$\begin{bmatrix} 0 & 0 & 0 & \cdots & 0 \\ p_1 & -\lambda_0 & 0 & \cdots & 0 \\ 0 & p_2 & -\lambda_0 & \cdots & 0 \\ \vdots & \vdots & & & \vdots \\ 0 & 0 & 0 & \cdots & -\lambda_0 \end{bmatrix}$$

from which it is clear that $\mathbf{N}_0$ given by $(B - \lambda_0 I)\mathbf{N}_0$ can be taken to have all non-negative components. For example, if $m = 4$, the components can be taken to be

$$n_1 = \lambda_0^3, \quad n_2 = \lambda_0^2 p_1, \quad n_3 = \lambda_0 p_1 p_2, \quad n_4 = p_1 p_2 p_3,$$

∎

Problems 15.2

1. Find the stable age distribution for the following Leslie matrix:

$$B = \begin{bmatrix} 0 & 3 & 14 & 36 \\ 1/2 & 0 & 0 & 0 \\ 0 & 1/4 & 0 & 0 \\ 0 & 0 & 2/3 & 0 \end{bmatrix}.$$

2. Find the inverse of B.

3. Let $b_1 = b_2 = b_3 = 0$, $b_4 = b$, and $p_1 = p_2 = p_3 = p$; find the general forms of $N(t)$, λ_0, and N_0.

15.3 Leontief Input–Output Analysis

Leontief input–output analysis is a technique for analyzing a portion of an economy. The flows of goods or services into and out of a particular industry are combined with those of the other industries to obtain a model of the entire economy. Since flows are studied, it is natural to interpret this as a compartmental model.

We have already seen an example of the Leontief theory treated from the point of view of Markov chains (Chapter 8, Example 10). However, it is more natural to treat this subject in terms of a continuous- rather than discrete-time model.

We shall only discuss the closed model of an economy with no exchange of goods or services with the outside world. In this model, the various industries into which the economy is partitioned form the compartments; the output of each industry is consumed entirely by it and the others. Also, the supplies for each industry come entirely from the other industries (or from itself).

The flow from industry i to industry j is assumed to be donor controlled, but in terms of price rather than the goods or services. More precisely, it is of the form $a_{ij}p_i$, where p_i is the price of the total output of industry i. The coefficient a_{ij} is the fraction of the total output of industry i purchased by industry j per unit time. The model then becomes a standard compartmental model with price as the material whose flows are followed. It should be observed that the price may change (and usually will) even though the total output of goods and services remains the same.

The differential equation is

$$dp/dt = Ap,$$

where **p** is the price vector. Since it is a closed system, the matrix A is singular and its columns add up to zero (see Part IV). If the digraph of the model is strongly connected, then the solution

$$\mathbf{p}(t) = \exp(At)\mathbf{p}(0)$$

converges to an equilibrium value in which each compartment has a positive level; that is, the industries can all coexist in equilibrium with a positive price for their outputs. In fact, this equilibrium solution is often the only aspect of the model considered.

If the digraph is not strongly connected, but is weakly connected, then any compartment in the vertex basis (see Chapter 3) will eventually empty out. The equilibrium solution will be one in which the corresponding industry will be paid nothing for its output. If the digraph is not weakly connected, then each of the weak components must be examined separately.

Example 15.1

We consider the same example that we considered previously in the Markov chain context (Chapter 8, Example 10). This consists of three industries corresponding to the work done by a carpenter, plumber, and electrician on each of their respective houses. The portion of the weekly output of each of the three on the three houses is given by the diagram in Figure 15.7. Thus, for example, the output of the carpenter purchased by the plumber is 0.4 of his weekly output. The differential equation is

$$dp/dt = \begin{bmatrix} -0.8 & 0.1 & 0.6 \\ 0.4 & -0.7 & 0.1 \\ 0.4 & 0.6 & -0.7 \end{bmatrix} \mathbf{p}.$$

The eigenvalues of the matrix are $\lambda_1 = 0$, $\lambda_{2,3} = -11 \pm i\sqrt{6}$. The eigenvector corresponding to λ_1 is

$$\mathbf{v}^T = [43, 32, 52].$$

Hence, $\mathbf{p}(t)$ tends toward some multiple of the eigenvector as t tends to ∞. If the three craftsmen are each paid \$400 for their weekly output initially, then their final wage will be

$$\mathbf{p}^T(\infty) = [405, 302, 491].$$

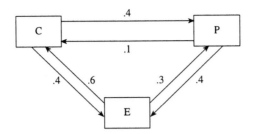

Figure 15.7: The diagram of a compartmental model for three industries.

It should be observed that their wages, i.e., prices, will not approach this value monotonically, but will have an oscillatory component because of the complex eigenvalues $\lambda_{2,3}$.

Problems 15.3

1. Find the closed-form solution of the example and determine how many weeks it will take before the wages are within 2% of their final value.

2. Suppose the house of the electrician requires no plumbing. Modify the compartmental model to conform to this. Then, find the final wage of each in this case.

16

Ecosystem Models

Compartmental models and their extensions are natural for studying ecosystems since flows of energy and nutrients (nitrates, phosphates, carbon, etc.) drive the system.

We have already seen some simple examples which, however, have not been quantitative. In this chapter, we present some additional examples for which plausible data have been obtained. They include models for dissolved oxygen in stream, for nutrient flow in forest growth, and for biomass flow in fisheries.

16.1 Dissolved Oxygen

The concentration of dissolved oxygen in streams and lakes is often considered an indication of the degree of pollution. In a nonpolluted body of water, a diverse community of green plants produces oxygen through photosynthesis. Oxygen is lost primarily through respiration of the plants and bacteria. In a polluted environment, the sources of oxygen are reduced and the consumers increased.

For a stationary body of water, the concentration is modeled by a three-compartment system as in Figure 16.1. The flow into compartment C has two components, one depending solely on the production and the other depending on the concentration. This leads to an equation of the form usually written as

$$\frac{dc}{dt} = k(c_s - c) + p - r \tag{16.1}$$

(see Odum (1956) or O'Connor and DiToro (1970)), where c_s denotes the saturation concentration, p the production of oxygen from algae, and r the respiration loss. The production occurs only during daylight hours and has

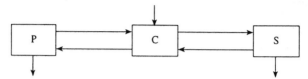

Figure 16.1: Diagram for compartmental model of dissolved oxygen (DO). The compartments correspond to DO concentration (C), plants (P), and bacteria (S).

the form

$$p(t) = p_m \sin \rho(t - t_s)\chi_{[0,\frac{\pi}{\rho}]}(t - t_s),\tag{16.2}$$

where p_m is the maximum production and the other terms are self-explanatory.

For a particular station on the Flint river, O'Connor and DiToro give the values $k = 4.2$ and $c_s = 8$ mg/L, $p_m = 35$ mg/L per day, and $r = 15$ m/L per day. Equation (16.1) has values

$$p(t) = 35 \sin 2\pi(t - 1/4)\chi_{[0,\frac{1}{2}]}(t - 1/4),$$

where t is measured in days. The initial concentration is assumed to be the saturation level $c(0) = c_s$. Equation (16.1) then becomes

$$\frac{dc}{dt} = 4.2(8 - c) + 35 \sin 2\pi(t - 1/4)\chi_{[0,\frac{1}{2}]}(t - 1/4) - 15$$

whose solution for the first day is

$$c(t) = 8e^{-4.2t} + e^{-4.2t}\int_0^t e^{04.2s}(18.6 + p(s))ds$$

$$= 8e^{-4.2t} + \frac{18.6}{4.2} - \frac{18.6}{4.2}e^{-4.2t}$$

$$+ 35e^{-4.2t}\int_{1/4}^t e^{4.2s}\sin 2\pi(s - \frac{1}{4})\chi_{[0,\frac{1}{2}]}(s)\frac{1}{4}ds$$

$$= \begin{cases} 3.6e^{-4.2t} + 4.4, & 0 \leq t \leq 1/4, \\ 3.6e^{-4.2t} + 8.3 + 2.6\sin 2\pi(t - 1/4) - 3.9\cos 2\pi(t - 1/4), \\ \qquad\qquad\qquad\qquad 1/4 \leq t \leq 3/4, \\ 3.6e^{-4.2t} + 12.2, & 3/4 \leq t \leq 1. \end{cases}$$

In subsequent days, the calculations are the same except that the initial condition is $c_0 = c(1)$.

This model does not take into account stream flow. To do so involves adding another couple of terms. Also, in order to study the effects, say, of pollution, there might be several monitoring stations. Thus, the compartmental model might have the form in Figure 16.2.

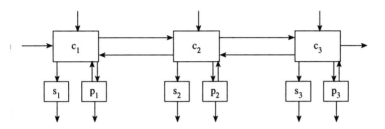

Figure 16.2: Diagram of a compartmental model of dissolved oxygen at three locations.

The flow is from compartment 1 to compartment 2 to compartment 3. The term that is added involves the change in concentration and the velocity of the stream. Thus, for the second compartment, the equation is of the form

$$\frac{dc_2}{dt} = k_2(c_1 - c_2) + p_2 - r_2 + (c_2 - c_3)\frac{v}{d_{23}} + (c_1 - c_2)\frac{v}{d_{12}}, \qquad (16.3)$$

where v is the stream velocity and d_{23} and d_{12} are the distances between stations. Similar equations can be written for c_1 and c_3. However, we shall not do so since their analysis, although still elementary, requires the solution of a system of coupled equations. In most of the literature, the differences $(c_2 - c_3)/d_{23}$ are replaced by the partial derivatives $\frac{\partial c_2}{\partial x}$, but this requires a knowledge of p_2 and r_2 as a function of x (see O'Connor and DiToro).

Problem 16.1

Suppose the maximum production p_m is reduced by 50%, but the other values are the same in the model of Figure 16.1. Find the new $c(t)$.

16.2   Forest Ecosystem

Most forest ecosystems are incredibly complex, particularly in the tropics. However, the models are often simplified to just four compartments: leaves, debris, soil, and wood. The flows are as indicated in Figure 16.3. The leaves fall and contribute to the debris, which, in turn, decomposes into soil. The nutrients from the soil are converted into wood and then leaves. Flows into and out of the system are primarily through the soil compartment. The flows could serve for any nutrient, but we shall concentrate on potassium flows and use data corresponding to it. The flows are assumed to be donor controlled, which is plausible for each of the compartments. We shall apply the same diagram to both a temperate and a tropical forest. But the flow rates are quite different in that the decomposition of the debris is an order of magnitude faster in the tropical forest. Typical values for a temperate forest (in units of g/ha yr) are $a_{12} = 50$, $a_{23} = 90$, $a_{34} = 60$, and $a_{41} = 50$. For a tropical forest, typical values might be $a_{12} = 90$, $a_{23} = 900$, $a_{34} = 60$, and $a_{41} = 10$.

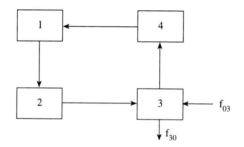

Figure 16.3: Flows of potassium between compartments of a forest: (1) leaves, (2) debris, (3) soil, and (4) wood.

For the moment, we assume the inputs are balanced by the outputs, i.e., $f_{03} = f_{30}$, so that the equation has the form

$$\frac{d\mathbf{x}}{dt} = A\mathbf{x}, \tag{16.4}$$

where A has the form

$$A = \begin{bmatrix} -a_{12} & 0 & 0 & a_{41} \\ a_{12} & -a_{23} & 0 & 0 \\ 0 & a_{23} & -a_{34} & 0 \\ 0 & 0 & a_{34} & -a_{41} \end{bmatrix}. \tag{16.5}$$

The associated digraph of this model is strongly connected; hence, Equation (16.4) has an equilibrium solution, all of whose components are positive. (Recall that such equilibrium solutions are the same for closed donor-controlled compartmental models as for regular Markov chains, as discussed in Chapter 12.) This equilibrium solution is easy to find because of the cyclic nature of A. In fact, the system $A\mathbf{x} = \mathbf{0}$ is equivalent to

$$A\mathbf{x} = \begin{bmatrix} -a_{12} & 0 & 0 & a_{41} \\ 0 & -a_{23} & 0 & a_{41} \\ 0 & 0 & -a_{34} & a_{41} \\ 0 & 0 & 0 & 0 \end{bmatrix} \mathbf{x} = \mathbf{0}.$$

from which it easily follows that $x_1 = 1/a_{12}$, $x_2 = 1/a_{23}$, $x_3 = 1/a_{34}$, and $x_4 = 1/a_{41}$ is an equilibrium solution. Thus, for the two sets of values considered, we find that for the temperature forest, $x_1 = 1$, $x_2 = 5/9$, $x_3 = 5/6$, $x_4 = 1$, where we have normalized the solution to make the wood compartment (4) level equal to 1. If we use the same normalization for the tropical forest, we find that $x_1 = 1/9$, $x_2 = 1/90$, and $x_3 = 1/6$, so that a far larger position of the nutrient is found in the wood in this latter case. Especially striking is the fact that only about 1% of the nutrients are found in the debris and about 10% in the soil. Hence, if you cut a tropical forest, you remove most of the nutrients, whereas in a temperate forest, you lose much less.

Another similar and perhaps more realistic model has been described in the work of Patten et al. (1976). The model follows the flow of nitrogen through a tropical forest with five compartments: (1) leaves, (2) debris, (3) filtrous roots, (4) soil, (5) wood. It is given in Figure 16.4.

Again, the inputs balance the outputs so that the system tends to an equilibrium solution of the differential equation which has the form

$$\frac{d\mathbf{x}}{dt} = A\mathbf{x} + \mathbf{b} \tag{16.6}$$

in contrast to (16.4). In the former case, the net flow to the outside was the $\mathbf{0}$ vector, but in this case, it is a vector which sums to 0. Thus, the solution of the equilibrium equation

$$A\mathbf{x} + \mathbf{b} = \mathbf{0} \tag{16.7}$$

is what we must find. Since both A and the augmented matrix $[A \mid b]$ have the same rank, there is a solution. For the values in Figure 16.4, we have

$$\begin{bmatrix} -1.05 & 0 & 0 & 0 & 0 \\ 1.05 & -1.27 & 0 & 0 & 0 \\ 0 & 1.27 & -4.01 & 0 & 0 \\ 0 & 0 & 2.88 & -.005 & 0 \\ 0 & 0 & 1.13 & .005 & -.098 \end{bmatrix} \mathbf{x} + \begin{bmatrix} 5.8 \\ 0 \\ -3.2 \\ -2.6 \\ 0 \end{bmatrix} = \mathbf{0}.$$

whose solution is

$$\mathbf{x}^T = [-3.05, -2.52, 0, 520, 26.5] + C[3.82, 316, 1, 576, 41.0].$$

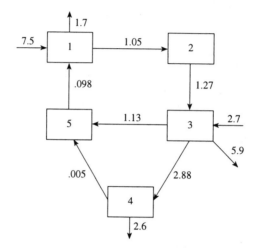

Figure 16.4: A compartmental model of nitrogen flow in a tropical forest: The weights on the arrows between compartments are donor-controlled flow rates; the inputs and outputs are assumed constant.

The constant C depends on the total amount of nitrogen in the system at the start. Notice that if C is too small, the equilibrium solution could have negative values, which means the system will collapse in a finite time.

Problem 16.2

Find the eigenvalues and eigenvectors of the matrix A for both the temperate and tropical forests (by using Maple or other computer program). Now, suppose some fertilizer is introduced (1 kg/ha). How long before half is absorbed in the wood compartment in each case?

16.3   Food Webs

If one follows the flow of energy (rather than nutrients) through an ecosystem, it proceeds from the primary producers (plants) to herbivores to carnivores and eventually it decomposes. This describes a *food chain* whose digraph is a simple path.

Very few realistic ecosystems have such simple chain structures. However, the flow of energy can often be modeled realistically by a *food web*. The digraph of such a structure is acyclic, with the energy flowing from the primary producers to the apex predators or to the "outside." We shall consider the latter case in which the outside is adjoined to the system as another compartment. In this case, it is clear that everything must ultimately end up in this last compartment. The only question is how long it will take if an impulse of food is supplied to any of the primary producers.

Let us assume that the compartments are ordered according to some measure of tropic level, i.e., some integer-valued function of the compartments such that a predator always has a higher level than its direct or indirect prey. It is easy to show such a measure exists for all acyclic digraphs. With this ordering, the matrix of the compartmental model is lower triangular and has the form

$$A = \begin{bmatrix} -a_{11} & 0 & \cdots & 0 \\ a_{12} & -a_{22} & \cdots & 0 \\ a_{13} & a_{23} & \cdots & 0 \\ \vdots & \vdots & \cdots & 0 \\ a_{1n} & a_{2n} & \cdots & 0 \end{bmatrix}. \tag{16.8}$$

For example, a system with two primary producers, two herbivores and

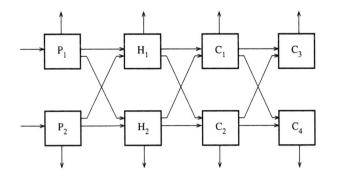

Figure 16.5: A compartmental model of a food web with primary producers (P_1, P_2), herbivores (H_1, H_2), and carnivores (C_1, C_2, C_3, C_4).

four carnivores might be as in Figure 16.5. The flows are taken to be donor controlled except for the inputs to compartments P_1 and P_2. We model this with a discrete-time model, which we can treat as a Markov chain. The appropriate type in this case is an absorbing chain with the outside as the absorbing compartment. It has the form

$$P = \begin{bmatrix} Q & 0 \\ R & 1 \end{bmatrix},$$

where $Q = (I + hA_{11})^T$ is an $(n-1) \times (n-1)$ submatrix. From the theory of such chains, the expected time to absorption is given by the sum of those rows of $(I - Q)^{-1} = N$ corresponding to inputs. This time is units of h by which N must be multiplied to obtain the time in the original units.

The matrix hN may be found directly by calculating the inverse of $I - Q = -hA_{11}^T$. In this case of a simple food chain, it is given by

$$hN = h(I - Q)^{-1} = -A_{11}^{-T} = \begin{bmatrix} 1/a_1 & 1/a_2 & \cdots & 1/a_{n-1} \\ 0 & 1/a_2 & \cdots & 1/a_{n-1} \\ \vdots & \vdots & \cdots & \cdots \\ 0 & 0 & \cdots & 1/a_{n-1} \end{bmatrix},$$

where a_i is the flow rate from compartment i to compartment $i+1$. The expected time to absorption is thus

$$E = \sum_{i=1}^{n-1} a_i^{-1}.$$

This, of course, is exactly what one would suspect.

Problem 16.3

Construct a compartmental model of a food web with one primary producer, two herbivores, and one carnivore. Find the compartmental matrix for the case of equal flow rates. Then, imitate the procedure used to find the expected time to absorption.

17

Fisheries Models

Ecosystem models are generally used to try to understand and predict the behavior of the system. However, in the case of fisheries models, they are used for more than this. They are used to try to determine the optimum harvest and, for management of fisheries, to establish the allowable catch of the various species.

The management of renewable resources such as fisheries has usually been based on the concept of *maximum sustainable yield*(MSY). It assumes that either too much or too little fishing or hunting or gathering would reduce the amount obtained in the long run. Too much would reduce the population excessively, whereas too little would result in a small harvest. Thus, the amount of the harvest under equilibrium conditions (y_e, equilibrium yield) is assumed to be a convex function of the harvest effort f (see Figure 17.1) with a maximum attained at some intermediate value between smaller and larger efforts. The same equilibrium yield can also be expressed as a function of population biomass x which would have the same functional form.

One approach to fisheries management based on such a concept is the detailed model of Beverton and Holt (1957). It is applicable to fisheries for which a great number of data regarding mortality, age structure, and growth rate are available. In this chapter, we shall not consider this approach but will confine ourselves to the *simple* or *surplus yield* models proposed initially by Schaefer (1954). These models, while not as accurate, often are the only tool available, since they require only catch and effort data for their implementation.

We shall briefly review Schaefer's approach and then present some criticisms that have been directed against it. These criticisms have led to modifications in his approach designed to correct and extend it. The major portion of this chapter is devoted to describing and comparing these modifications.

17.1 Logistic Equation Approach

The Schaefer model is based on a one-compartmental model in which the flows in and out are quadratic functions of the level. This has already been covered in Chapter 13 (Section 13.2.2) and leads to the logistic equation. It is usually written in terms of the biomass x of a "stock" of fish, i.e., a set composed of fish from a single species that live together and reproduce together:

$$\frac{1}{x}\frac{dx}{dt} = r\left[1 - \frac{x}{x_\infty}\right], \qquad (17.1)$$

where r is the intrinsic growth rate and x_∞ is the carrying capacity of the environment. Its solution is

$$x(t) = \left[x_\infty^{-1} - \left(x_\infty^{-1} - x_o^{-1}\right)e^{-rt}\right]^{-1}, \qquad (17.2)$$

a monotonically increasing function of time t which starts from the initial level x_o and approaches x_∞. It was used by Schaefer (1954) to model a fishery by adding a term corresponding to the fishing mortality rate:

$$\frac{1}{x}\frac{dx}{dt} = r\left[1 - \frac{x}{x_\infty}\right] - qf. \qquad (17.3)$$

Here, f is the fishing effort and q is the catchability coefficient. It assumes that the instantaneous fishing mortality rate is proportional to the effort, e.g., the number of boat-days fished.

Schaefer used this model to construct an equilibrium yield curve and then to calculate the maximum sustainable yield based on it. The annual yield or catch is obtained by integrating the rate of fishing mortality times the biomass over a year. If during this year the fishing mortality exactly balances the growth, i.e., if $\frac{dx}{dt} = 0$, then the yield becomes the equilibrium yield and is given by

$$y_e = qfx = r\left[1 - \frac{x}{x_\infty}\right]x. \qquad (17.4)$$

Hence, one can calculate MSY to be the maximum of y_e:

$$\text{MSY} = rx_\infty/2. \qquad (17.5)$$

In order to regulate the fishery, it is better to express this in terms of effort by solving (17.4) for x in terms of f and then replacing f in the yield expression to get

$$y_e = qfx_\infty\left[1 - \frac{qf}{r}\right]. \qquad (17.6)$$

The graph of y_e vs. x and y_e vs. f are therefore similar and are given in Figure 17.1. Thus, if regulation of the fishering is to be based on catch, the maximum value would be $rx_\infty/2$, whereas if it is to be based on effort the optimum value would be $r/2q$. In either case, all that is needed is an estimate of the parameters q, r, and x_∞. Schaefer based his estimates of these parameters on the catch and effort data only. Such data are available readily for many fisheries; the detailed data needed for other approaches are not.

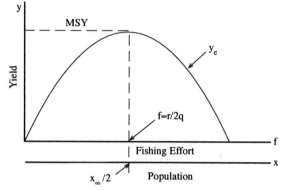

Figure 17.1: Plot of equilibrium yield (y_e) vs. fishing effort (f) and population biomass (x) for the Schaefer model. The maximum sustainabile yield (MSY) occurs at $f = r/2q$ or at $x = x_\infty/2$.

He assumed, as does the model implicitly, that the population biomass is proportional to U, the catch per unit effort under equilibrium conditions. The constant of proportionality is exactly the catchability coefficient q. The equation satisfied by this U is obtained by dividing (17.6) by f:

$$U = qx_\infty \left[1 - \frac{q}{r}f \right] .$$

Thus, the data can be plotted on the $U - f$ plane and a straight line fitted to it, say,

$$U = \hat{a} - \hat{b}f.$$

The fitted parameters $\hat{a}$ and $\hat{b}$ then become estimates of qx_∞ and q^2x_∞/r, respectively. These, in turn, may be used to estimate the optimum effort

$$f_{\text{opt}} = \frac{r}{2q} \approx \frac{\hat{a}}{2\hat{b}}.$$

Similarly, an estimate for the MSY may be obtained:

$$\text{MSY} \approx \frac{\hat{a}^2}{4\hat{b}}.$$

Schaefer did this for yellow fin tuna and halibut in the Pacific and the data seemed to fall pretty close to the straight line.

There is another problem with this fitting method even for good data—it assumes that the population is approximately in equilibrium. This has rarely been the case in recent years since most fisheries have experienced a rapid increase in effort accompanied by a decline in biomass. Even if all of the data fell on a straight line, this would not be the line of equilibrium; in a declining fishery it would lie above it, often far above it.

There are actually two problems here: one concerned with the fitting of data and the other with the analysis. They constitute the first two criticisms of Schaefer's approach; others of which follow.

(A) The plot of U vs. f often consists of data from a fishery not in equilibrium or even not in periodic equilibrium. The straight-line fit to such data then is not the equilibrium line.

(B) Regulation based on MSY is inappropriate for a fishery not in equilibrium; in fact, it could lead to a catastrophe if the population is small.

(C) Equations other than the logistic should be used since they more accurately reflect biological behavior.

(D) No stock lives in a vacuum; interactions with other species and the environment should be included in any realistic analysis.

(E) A fishery is an economic enterprise and should be analyzed as such. Economic yield rather than biomass yield should be maximized; since future catches are worth less than present catches, they should be accordingly discounted.

(F) The world is not the regular place of deterministic differential equations but rather full of variability, so a stochastic variable which accounts for this should appear in the equation.

Some of these objections have no good answers. Many alternatives for parameter estimation have been proposed, but fisheries data frequently have so much noise that no reasonable method works. Then other models are perhaps better.

We shall discuss a few of the alternatives proposed for B, C, and D. We begin with an alternative regulation procedure designed to meet objection B.

17.2   Nonequilibrium Yield

The regulation based on MSY is particularly dangerous since the value of yield that corresponds to it is a critical value; that is, a constant yield strategy above the MSY leads to annihilation of the stock in finite time, whereas below the MSY level, it leads to a positive asymptotic value of population biomass. The presence of such a jump in behavior is called a catastrophe.

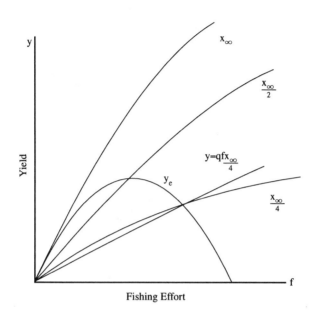

Figure 17.2: Plot of equilibrium yield (y_e) and annual attainable yield for three different values of initial population. The straight line $y = qfx_\infty/4$ is the approximation to the annual attainable yield for initial population $x_\infty/4$ given by Equation (17.8).

In order to avoid this difficulty, some method of regulation based on the fact that the fishery is rarely in equilibrium must be used. The attainable annual yield, in the case of nonequilibrium, may be found by integrating the catch rate qfx over the nth year. The closed-form solution of $x(t)$ given by (17.2) (with the initial value at the start of this year) is substituted for x in this integral to obtain

$$y_n = \int_n^{n+1} qfx(t)dt = qf \int_n^{n+1} \frac{r_1 x_\infty \, dt}{r - [r - r_1 x_\infty/x_n]e^{-r_1(t-n)}}, \qquad (17.7)$$

where $r_1 = r - qf$, the intrinsic growth rate in the presence of fishing. Hence, we have

$$y_n = \frac{qfx_\infty}{r} \ln \left[e^{r_1(t-n)} - 1 + \frac{r_1 x_\infty}{r x_n} \right]_n^{n+1}$$

$$= \frac{qfx_\infty}{r} \ln \left[\frac{e^{r_1} - 1 + \frac{r_1 x_\infty}{r x_n}}{\frac{r_1 x_\infty}{r x_n}} \right]$$

$$\approx \frac{qfx_\infty(e^{r_1} - 1)}{rr_1 x_\infty / rx_n} \approx qfx_n. \tag{17.8}$$

Hence, this annual yield is approximately a straight line with slope qx_n in the y-f plane. (See Figure 17.2).

A number of different regulation strategies based on nonequilibrium yield are possible. A safe procedure is to regulate effort instead of catch. However, in many cases, there is no regulation, and the fishery is allowed *open access*. Still, in this case, some control is possible through taxation or subsidies. The cost of fishing is assumed proportional to effort and the return is proportional to yield. If the price is subsidized or taxed in such a way that at MSY, the effort cost equals the yield price, then the MSY point will be approached. Graphically, it is as in Figure 17.3. This overestima-

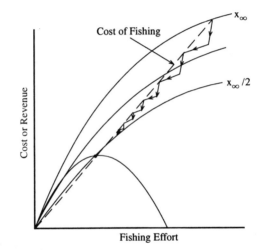

Figure 17.3: Trajectory of yield and effort under an open-access strategy. The cost of fishing is assumed to be a linear function of effort adjusted to pass through the point (f_{opt}, MSY) at the peak of the equilibrium yield curve. At each level of the population, the effort is adjusted to bring revenue and cost into equilibrium. However, this approach has rarely been used. Rather, regulation based on an incorrect value of MSY has been used.

tion of MSY and resulting decline seems to have happened in a number of the world's fisheries. In particular, the Atlantic mackerel fishery in the Northwest Atlantic seems to have been subject to this sort of error. The data for this fishery are shown in Figure 17.4 (see Walter (1976)).

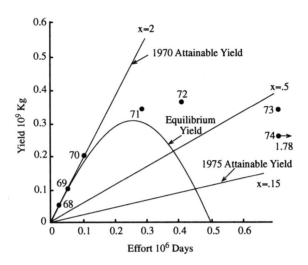

Figure 17.4: Equilibrium yield and annual attainable yield curves for a stock of Atlantic mackerel (*Scomber scombrus*). The virgin population is estimated at $x = 2.4 \times 10^9$ kg. The straight lines are approximations to the annual attainable yield for 1970 ($x = 2 \times 10^9$ kg $= 5/6x_\infty$) for an intermediate value, and for 1975 ($P = 0.15 \times 10^9$ kg $= 0.06x_\infty$). High effort produced high yields—close to MSY—even with a population far below $x_\infty/2$ (modified from Walter (1976)).

Other equations which also correspond to a compartmental model with one compartment have been proposed. They involve changing the functional form of the flow rates. Among them is the equation

$$\frac{1}{x}\frac{dx}{dt} = r\left[1 - \frac{\ln x}{\ln x_\infty}\right],$$

which was proposed by Fox (1970). This differs from the Schaefer equation in that the equilibrium yield is always positive for any positive level of fishing effort. It is given by

$$y_e = qfx_\infty^{1-qf/r}.$$

A general form including an additional parameter was proposed by Pella and Tomlinson (1969). It includes both Fox's and Schaefer's equation as special cases; it is

$$\frac{1}{x}\frac{dx}{dt} = r\left[1 - \left[\frac{x}{x_\infty}\right]^{m-1}\right].$$

By an appropriate choice of m, the yield curve can be assumed to be skewed in either direction. Otherwise, the analysis is much like that of Schaefer.

17.3   A Multispecies Model

The desirability of using a multispecies approach is clear. Many of the commercially fished species inhabit the same region and compete for the same food. The reduction of one stock may make more resources available to another and thus allow its sustainable yield to be increase.

To construct a model which takes these interactions into account, we use a compartmental model where the flow rates are similar to the one-compartment case which led to the logistic equation. For two compartments, the diagram is as in Figure 17.5.

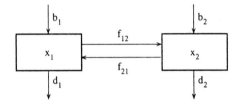

Figure 17.5: A two-compartment model for a two species fishery. Each of the flows from and to the outside (b_1, d_1, b_2, d_2) are quadratic functions of the levels of the associated compartment. The flows between compartments are also quadratic functions of both the donor and recipient compartments.

By collecting like terms, the equations of the two-compartmental model can be put into the forms

$$\frac{1}{x_1}\frac{dx_1}{dt} = b_1 - a_{11}x_1 - a_{12}x_2,$$

$$\frac{1}{x_2}\frac{dx_2}{dt} = b_2 - a_{21}x_1 - a_{22}x_2. \qquad (17.9)$$

These are the Lotka–Volterra competition equations. They may be extended to any number of species in the obvious way.

The analog to the Schaefer equation is obtained by adding a fishing mortality term to the Lotka–Volterra equation to obtain

$$\frac{d\ln\mathbf{x}}{dt} = \mathbf{b} - A\mathbf{x} - Q\mathbf{f}, \qquad (17.10)$$

where A and Q are nonsingular matrices and $\mathbf{b}$, $\mathbf{x}$, and $\mathbf{f}$ are column vectors.

The matrix A is the interaction matrix. If all the species are in competition, its elements will be non-negative. The element a_{ij} will correspond to the reduction in growth rate of the ith species due to the presence of a biomass of x_j of the jth species. However, if food fish are included, the corresponding elements would be negative. Q is the matrix of catchability coefficients. The components of the column vector $\mathbf{b}$ are quotients $r_i/x_{i\infty}$ of the intrinsic growth rate over the carrying capacity for the ith species. The components of the x and f are the biomasses and directed fishing efforts of each species. For two species, (17.10) has the detailed forms

$$\frac{1}{x_1}\frac{dx_1}{dt} = b_1 - a_{11}x_1 - a_{12}x_2 - q_{11}f_1 - q_{12}f_2,$$

$$\frac{1}{x_2}\frac{dx_2}{dt} = b_2 - a_{21}x_1 - a_{22}x_2 - q_{21}f_1 - q_{22}f_2. \tag{17.11}$$

The analysis of (17.10) proceeds in much the same way as the Schaefer equation to determine MSY. The equilibrium yield for all species is given by

$$y_e = \mathbf{x}^T Q \mathbf{f}, \tag{17.12}$$

which assigns the same weight to each species. However, if the species vary in economic importance, it may be desirable to weigh them differently. Then, the economic yield becomes

$$y_e = \mathbf{x}^T E Q \mathbf{f}, \tag{17.13}$$

where E is a diagonal matrix of weights (dollars/lb., etc.). Under equilibrium conditions, $\mathbf{x} = A^{-1}(\mathbf{b} - Q\mathbf{f})$ and, hence,

$$y_e = (\mathbf{b} - Q\mathbf{f})^T A^{-1^T} E Q \mathbf{f}.$$

This is a quadratic function in $\mathbf{f}$ whose maximum is attained at that value where the gradient $\nabla y_e = \mathbf{0}^T$. But since $\nabla \mathbf{f} = I$ and since $\nabla \mathbf{v}^T \mathbf{w} = ((\nabla \mathbf{v}^T)\mathbf{w})^T + \mathbf{v}^T \nabla \mathbf{w}$, we have, by taking $\mathbf{v} = \mathbf{b} - Q\mathbf{f}$ and $\mathbf{w} = A^{-1^T} E Q \mathbf{f}$,

$$\nabla y_c = \left[\frac{\partial y_e}{\partial f_1}\frac{\partial y_e}{\partial f_2}\cdots\frac{\partial y_e}{\partial f_n}\right]$$
$$= \left(-(QI)^T A^{-1^T} E Q \mathbf{f}\right)^T + (b - Q\mathbf{f})^T A^{-1^T} E Q I$$
$$= -\mathbf{f}^T Q^T E A^{-1} Q - \mathbf{f}^T Q^T A^{-1^T} E Q + \mathbf{b}^T A^{-1^T} E Q = \mathbf{0}^T.$$
$$\tag{17.14}$$

This may be solved for $\mathbf{f}^T$ to obtain

$$\mathbf{f}^T = \mathbf{b}^T A^{-1^T} E Q \left[Q^T A^{-1^T} E Q + Q^T E A^{-1} Q\right]^{-1}$$
$$= \mathbf{b}^T A^{-1^T} E \left[E A^{-1} + (E A^{-1})^T\right]^{-1} Q^{-1^T}, \tag{17.15}$$

provided these inverses exist. Hence, by taking transposes, the effort vector is found to be

$$\mathbf{f}_{\text{opt}} = Q^{-1} \left[EA^{-1} + \left(EA^{-1} \right)^T \right]^{-1} \left(AE^{-1} \right)^{-1} \mathbf{b}$$

$$= Q^{-1} E^{-1} A^T C^{-1} \mathbf{b}, \qquad (17.16)$$

where $C = AE^{-1} + E^{-1} A^T$. If each of the components of $\mathbf{f}_{\text{opt}}$ are positive, these values correspond to real efforts. If some are negative, the corresponding components are replaced by zeros and the optimum recalculated. The component values will not be the same in this case. In the former case, the maximum yield can be found by substitution to be

$$y_{\text{max}} = \left(\mathbf{b} - E^{-1} A^T C^{-1} \mathbf{b} \right)^T A^{-1T} E E^{-1} A^T C^{-1} \mathbf{b}$$

$$= \mathbf{b}^T \left(I - E^{-1} A^T C^{-1} \right)^T C^{-1} \mathbf{b}$$

$$= \mathbf{b}^T C^{-1} E^{-1} A^T C^{-1} \mathbf{b}. \qquad (17.17)$$

When A is symmetric, i.e., when the effect on the growth of the jth population by the ith population is the same as the effect on the ith by the jth, the equations are simpler. They are

$$\mathbf{f}_{\text{opt}} = 1/2 Q^{-1} \mathbf{b}, \qquad (17.18)$$

$$y_{\text{max}} = 1/4 \mathbf{b}^T E A^{-1} \mathbf{b}. \qquad (17.19)$$

The most difficult part is estimating the parameters. Straightforward least squares often leads to values which are negative when the proper interpretation of the model requires them to be positive.

Clearly, other methods of establishing the coefficients must be used—perhaps a combination of knowledge of biological interactions together with data fitting. This was attempted by Pope and Harris (1976) in studying stocks of South African pilchard and anchovy. The equations they obtained were

$$\frac{d \ln \mathbf{x}}{dt} = \begin{bmatrix} 0.43 \\ 1.10 \end{bmatrix} - 10^{-3} \begin{bmatrix} 0.14 & 0.14 \\ 0.10 & 0.5 \end{bmatrix} \mathbf{x} - Q\mathbf{f}.$$

The MSY for the two stocks together was found to be about 340,000 tons, whereas the individual MSYs, assuming no interactions, together totaled about 620,000 tons. This illustrates the importance of including interaction when there is a biologically valid reason for doing so.

Problem 17.3

Show the MSY for the two stocks in the paragraph above are correct.

17.4 An Ecosystem Fisheries Model

In this section, we attempt to construct a compartmental model of a real fishery, Georges Bank in the Northwest Atlantic off the coast of New England. This is a traditional fishing ground whose production declined sharply in the 1970s and has recently almost collapsed. We shall use available data (Maurer, 1975) to construct a food web digraph. This data consist of stomach contents of 10 species of finfish. The food consists of 7 of the 10 species as well as squid and other invertebrates.

Once we have the food web, we convert it into a compartmental model by incorporating additional data on the biomass of each species, its consumption to biomass ratio, and its growth efficiency. We then analyze this model by converting it to a Markov chain. This enables us to test the effect of various fishing strategies.

The starting point is the predator-prey matrix (Maurer, 1975) derived from a summary of stomach contents of 10 predators. These were taken from a region which included, but was not restricted to, Georges Bank during 1969-72. Only those prey of more than 100 g total per 10^6 of predator are included (Table 17.1).

This matrix is then used to construct a food web, as shown in Figure 17.6 where the abbreviations are self-explanatory. Each path represents the existence of more than 1 g of the donor compartment per kg of the recipient found in the latter's stomach.

In order to convert the food web digraph we use the data in Table 17.2 as well as the stomach content data. We assume the flows are linear and donor controlled and that material is biomass. Of course a gram of prey is not converted into a gram of predator but must be multiplied by a factor, the consumption to biomass ratio P_i of the table. Even this is not converted to growth since a certain portion of the food is used for metabolism. That is $1 - g_i$, where g_i is the growth efficiency in the table.

The biomass estimates in the table are those found in Clark and Brown (1977) for the years 1972-1974 on Georges Bank; the other data in the table are from Grosslein, Langton and Sissenwine (1978).

Table 17.1: Marine ecosystem model. Predator–prey relationships expressed as grams of prey per kilogram of predator (Maurer, 1975).

	Predators									
Prey	C	Ha	R	YT	OF_l	He	M	P	SH	OF
Cod	–	–	–	–	–	–	–	–	–	–
Haddock	–	–	–	–	–	–	–	–	–	–
Redfish	0.4	–	–	–	–	–	–	–	–	0.2
Yellowtail	0.3	–	–	–	–	–	–	–	–	–
Other flatfish[a]	0.2	–	–	–	–	–	–	–	–	0.3
Herring	2.5	–	–	–	–	–	–	–	1.3	–
Mackerel	0.4	–	–	–	–	–	–	–	2.4	1.3
Pollock	–	–	–	–	–	–	–	–	–	–
Silver hake	0.1	–	–	–	0.2	–	–	0.2	0.3	0.3
Other finfish[b]	7.6	–	–	–	–	–	–	4.4	5.5	6.7
Squid	–	–	–	–	0.1	–	–	–	0.3	1.2
Other invertebrates	5.3	2.8	4.5	2.0	2.8	2.7	2.2	10.4	3.4	3.9
Total	16.8	2.8	4.5	2.0	3.1	2.7	2.2	15.0	13.2	13.9

[a]This category includes five additional species.
[b]This predator category includes 15 other finfish; the prey category includes unidentifiable fish of all species, and fish eggs as well.

Table 17.2: Estimates of biomass, growth efficiency, and consumption to biomass ratio for each of the species considered on Georges Bank.

Comp. #	9	8	7	6	5	4	3	2
Species	C	SH	P	OF	[Ha, OF_e, YT, R]	He	M	Sq
Biomass X_i (10^5 MT)	2	3	0.2	3	1.3	4	13	6
Growth eff. g_i	0.19	0.12	0.11	0.11	0.11	0.06	0.10	0.23
Consum./ biomass P_i	3.2	5.0	4.1	4.1	4.0	4.5	4.4	7.0

The consumption rate by the ith compartment of the jth compartment, since we are assuming donor control, is given by $a_{ji}x_j(t)$.

We assume the system is in approximate equilibrium during the reference period since

$$a_{ji}\bar{x}_j = r_{ij}P_i\bar{x}_i,$$

where $\bar{x}_j$ is the mean biomass during the reference period and r_{ij} is the proportion from the jth compartment in the stomach of the ith compartment. This gives us an estimate of flow rates between compartments; $(1-g_i)\bar{x}_i$ is the flow to the outside of the system. The diagram of the compartmental model is shown in Figure 17.7. The four species R, XT, OF_l, and H_a are combined into a single compartment (#5) since they are all predators of

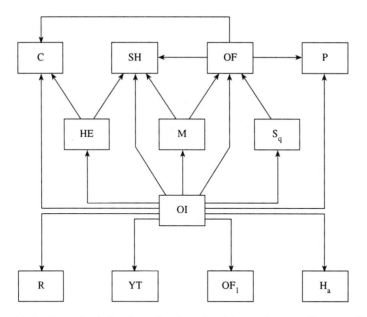

Figure 17.6: Principal food paths for the 12 species on Georges Bank. Only those paths corresponding to more than 1 gram of food per kilogram of predator are shown. The flow from OI to S_q is inferred since stomach content data for the latter were unavailable.

OI and do not serve as prey for others.

One thing missing is the estimate of the biomass of other invertebrates which drives the whole system. We take it to be 10 times the total fin fish biomass.

The differential equation $\frac{d\mathbf{x}}{dt} = A\mathbf{x}$ of this model has the matrix

$$A = 10^{-2} \begin{bmatrix} -31 & 0 & 0 & 0 & 0 & 0 & 0 & 0 & 0 \\ 0 & -95 & 0 & 0 & 0 & 0 & 0 & 0 & 0 \\ 17 & 0 & -120 & 0 & 0 & 0 & 0 & 0 & 0 \\ 7 & 0 & 0 & -155 & 0 & 0 & 0 & 0 & 0 \\ 2 & 0 & 0 & 0 & -89 & 0 & 0 & 0 & 0 \\ 3 & 18 & 9 & 0 & 0 & -402 & 0 & 0 & 0 \\ 0 & 0 & 0 & 0 & 0 & 8 & -89 & 0 & 0 \\ 1 & 0 & 21 & 37 & 0 & 208 & 0 & -88 & 0 \\ 1 & 0 & 0 & 24 & 0 & 97 & 0 & 0 & -81 \end{bmatrix}.$$

Since this is a triangular matrix, all the eigenvalues are the numbers on the main diagonal. Since they are all negative, the population of all species will decrease to zero. This is not unexpected since we have no inputs and eventually everything will be eaten up. We could guess at an input into compartment 1, but this would be less accurate than the figures we have.

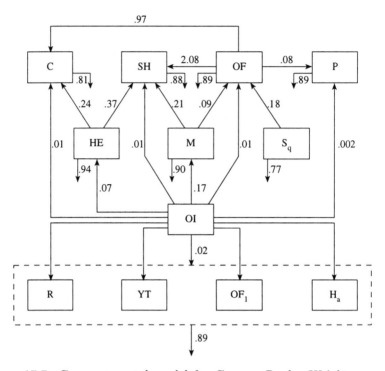

Figure 17.7: Compartmental model for Georges Bank. Weights on arcs correspond to donor-controlled flow in years^{-1}. The weights on arrows out of each compartment with no terminal end correspond to that proportion of incoming biomass which is not converted into that species biomass.

A more useful approach would be to consider an approach based on Markov chains. Compartments 5, 7, 8, and 9 are absorbing states if we exclude the excretion. However, we need to add another state corresponding to outside, which absorbs the biomass lost to metabolism, etc. Then, this new state becomes the only absorbing state. The transition matrix of the form

$$P = \begin{bmatrix} I & O \\ R & Q \end{bmatrix}$$

would have Q given by

$$Q = (I + hA)^T,$$

where h is an appropriate time step (say $h = 0.1$). Then, $(I - Q)^{-1}$ gives the expected time to absorption (Chapter 11).

18

Drug Kinetics

Drug kinetics have already been briefly discussed in Chapter 13 and an example given. However, there are many other examples in the literature. The interested reader is referred to the books by Gibaldi and Perrier (1982), Godfrey (1983), or Jacquez (1996) and their references. In this section, we take up a few additional examples.

In modeling drug kinetics in biomedical systems, we are usually interested in a tracer experiment. The underlying system may be nonlinear, but the introduction of a small quantity of the tracer will perturb the system only slightly. If this underlying system is close to equilibrium, then the level of the tracer can be adequately described by linear equations. In this chapter, we present four typical examples.

18.1 Bilirubin Metabolism (Simple)

Various types of liver diseases have been studied with the aid of compartmental models of Bilirubin metabolism (Anderson, 1983). An excess of Bilirubin causes the yellow jaundice associated with hepatitis.

The simplest case is a four-compartment system consisting of Bilirubin in the blood, liver, urine, and feces. The flows are as indicated in Figure 18.1. A tracer dose D is injected into the B compartment and levels measured over time in compartment B, U, and F. The compartmental model has equations

$$\mathbf{x}' = A\mathbf{x},$$

$$\mathbf{x}(0) = [D \quad 0 \quad 0 \quad 0]^T,$$

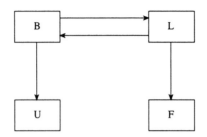

Figure 18.1: Bilirubin flows.

where

$$A = \begin{bmatrix} -a_{12} - a_{13} & a_{21} & 0 & 0 \\ a_{12} & -a_{21} - a_{24} & 0 & 0 \\ a_{13} & 0 & 0 & 0 \\ 0 & a_{24} & 0 & 0 \end{bmatrix}.$$

This matrix clearly has two negative eigenvalues and a double eigenvalue at 0. Hence, the solution to the differential equation is

$$\mathbf{x}(t) = e^{At}\mathbf{x}(0), \tag{18.1}$$

where $\mathbf{x}(0) = \begin{bmatrix} D & 0 & 0 & 0 \end{bmatrix}^T$.

A typical question confronting the experimenter is how to determine the flow coefficients $\{a_{ij}\}$ from a knowledge of the measurements. This is known as the *identifiability problem* and may not have a solution. It is addressed in more detail in Chapter 21. If we can solve this problem, then we can calculate the level in the liver compartment, which, in turn, may tell us something about the presence of disease.

In this example, the model is identifiable. If we know the derivatives of $\mathbf{x}(t)$ at $t = 0$, we can find A, since

$$\mathbf{x}'(t) = Ae^{At}\mathbf{x}(0)$$

and, therefore,

$$\mathbf{x}'(0) = A \begin{bmatrix} D \\ 0 \\ 0 \\ 0 \end{bmatrix} = D \begin{bmatrix} -a_{12} - a_{13} \\ a_{12} \\ a_{13} \\ 0 \end{bmatrix}.$$

Since we know $x_1(t)$ and $x_3(t)$, this gives us a_{13} and $-a_{12} - a_{13}$ and, hence, by addition, a_{12}. Similarly, we have

$$\mathbf{x}''(0) = A^2 \begin{bmatrix} D \\ 0 \\ 0 \\ 0 \end{bmatrix} = D \begin{bmatrix} (a_{12} + a_{13})^2 + a_{21}a_{12} \\ -(a_{12} + a_{13})a_{12} - (a_{21} + a_{24})a_{12} \\ -(a_{12} + a_{13})a_{13} \\ a_{12}a_{24} \end{bmatrix}.$$

From $x_4(t)$, we obtain $a_{12}a_{24}$, which gives us a_{24}, and from $x_1(t)$, we obtain a_{21} since we know a_{13} and a_{12}.

Hence, the system is identifiable, and we can find the level in each compartment for all time from (18.1).

Problem 18.1

For a particular patient, the values $a_{12} = 0.34$, $a_{13} = 0.02$, $a_{21} = 0.04$, and $a_{24} = 0.01$ were estimated (Anderson, 1983). Find the Bilirubin tracer in the patient's blood at any time if the initial dose D is 1.

18.2   Bilirubin Metabolism (Complex)

The model just discussed was essentially a two-compartment model since there were no exchanges between compartments 3 and 4 and other compartments. Another model involves splitting the liver compartment into two compartments—one the liver itself and the other the extra vascular pool of Bilirubin (see Anderson (1983) for more detail). It looks like Figure 18.2.

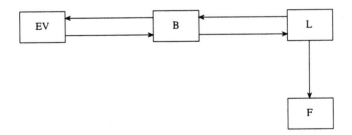

Figure 18.2: Flow of Bilirubin in the general model.

It is assumed that the discharges from the first and second compartment are negligible and that measurements are taken in compartments 1 and 4.

The matrix of the model now is of the form

$$
A = \begin{bmatrix}
-a_{12} - a_{13} & a_{21} & a_{31} & 0 \\
a_{12} & -a_{21} & 0 & 0 \\
a_{13} & 0 & -a_{31} - a_{34} & 0 \\
0 & 0 & a_{34} & 0
\end{bmatrix}.
$$

Again, an initial dose is injected in compartment 1. Now, however, from $\mathbf{x}'(0)$, we get only the sum of a_{12} and a_{13} and no matter how many powers of A we take, we cannot find all of the coefficients.

18.3   Lead Kinetics

Although lead is usually not introduced into the body intentionally, its flow through the body can be treated in the same way as drugs. Lead is introduced into the body in a number of ways. Lead plumbing and lead solder joints on other pipes cause small amounts of lead to be introduced into drinking water. The deterioration of lead-based paints introduces lead into the air and into dust and flakes where it is sometimes ingested by children. Before the advent of unleaded gasoline, automobile exhaust was a major source of environmental lead.

A three-compartment model of lead kinetics in the human body was studied by Batchelet et al. (1979). The model consisted of three compartments: blood, tissue, and skeleton as shown in Figure 18.3. We assume the

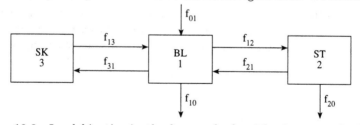

Figure 18.3: Lead kinetics in the human body. The input to the blood comes from both lungs and digestive tract. The output from the blood is mainly urine and from soft tissues, mainly sweat.

input is a constant b, and that the flows f_{ij}, as usual, are donor controlled and linear, $f_{ij} - a_{ij}x_i$. Then, the differential equation of the model has the form

$$
\frac{d\mathbf{x}}{dt} = \begin{bmatrix}
a_{11} & a_{21} & a_{31} \\
a_{12} & a_{22} & 0 \\
a_{13} & 0 & -a_{31}
\end{bmatrix} \mathbf{x} + \begin{bmatrix} b \\ 0 \\ 0 \end{bmatrix}
$$

$$
= A\mathbf{x} + \mathbf{b},
$$

where $a_{11} = -a_{12} - a_{13} - a_{10}$ and $a_{22} = -a_{21} - a_{20}$.

The digraph of the model is strongly connected and the matrix A has real eigenvalues. This can be shown directly in this case but also follows from the structure, which, in this case, is a mammillary system. (This is covered in Part IV.) These eigenvalues are all negative since they must be ≤ 0 and zero is not an eigenvalue. Thus, any solution to the equation converges to a stable equilibrium solution

$$\mathbf{x}_e = A^{-1}\mathbf{b}$$

as $t \to \infty$. All the components of $\mathbf{x}_e$ are positive and thus the lead is shared with all three compartments.

The flow rates for a particular individual were measured by Batchelet et al. and are the values given in the following matrix:

$$A = \begin{bmatrix} -0.036 & 0.0124 & 0.00035 \\ 0.011 & -0.0284 & 0 \\ 0.0039 & 0 & -0.00035 \end{bmatrix}.$$

This gave a solution to the equilibrium equation of

$$\mathbf{x}_e = A^{-1}\mathbf{b} = b \begin{bmatrix} 36.5 \\ 14.2 \\ 4057.1 \end{bmatrix}.$$

Thus, by far, the largest portion of the lead resides in the third compartment, the skeleton, ultimately. The eigenvalues are

$$\lambda_1 = -0.000031, \quad \lambda_2 = -0.020, \quad \lambda_3 = -0.045.$$

Once this large quantity of lead is in the skeleton, it is removed very slowly since the first eigenvalue is very small.

Problems 18.3

1. Use Maple to find the eigenvectors of the matrix A. (Find the null space of $A - \lambda I$.)

2. Expand $\mathbf{x}_e$ in terms of these eigenvectors. Then, find how long it would take for the lead in the skeleton to be reduced by 50% if no lead were ingested.

18.4   HIV Dynamics

The spread of AIDS has made it a significant health problem in much of the world in the last two decades. This disease, caused by the human immunodeficiency virus (HIV), has resisted efforts to find a cure. Part of this effort has involved mathematicians who have tried to find models both for the epidemiology and the biology of HIV infections. See Castillo-Chavez (1989).

We briefly present an approach for the latter involving compartmental models. It models the interaction between various types of T4-cells, which are the primary hosts of HIV.

Wu and Ding (1999) hypothesized the following compartmental model for HIV dynamics *in vivo*. They considered antiviral drug effects and possible infected cell and virus compartments. The concentration, and thus the compartment levels of these cells and viruses are denoted by T_m, T_s, T_l, T_p, V_I, and V_{NI}, where T_m, T_s, T_l, and T_p correspond to the mysterious infected cells , T_s to the long-lived infected cells, T_l to the latently infected cells, T_p to the productively infected cells, and V_I and V_{NI} to the infectious virus and noninfectious virus, respectively. The flow diagram describing the dynamics is given in Figure 18.4. Here, T denotes the quantity of uninfected

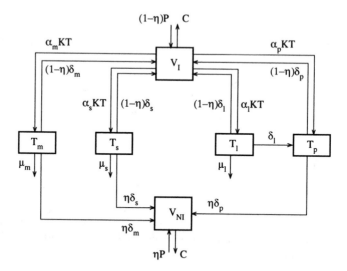

Figure 18.4: Flow diagram for the various types of cells and viruses in an HIV infection.

T cells, and α_m, K, α_s, α_l, α_p, δ_l, δ_m, δ_s, and δ_p the corresponding rates of infection. N is the average number of virions per cell, and μ_m, μ_s, and μ_l are the death rates of the infected cells. η is the proportion of noninfectious virus, C is the elimination rate of infectious and noninfectious virus, and, lastly, P is the virus production term or forcing function term assumed to be a constant average rate.

Thus, the diagram in Figure 18.4 gives rise to the system of differential equations:

$$\frac{d}{dt}T_m = \alpha_m KTV_I - (\delta_m + \mu_m)T_m,$$

$$\frac{d}{dt}T_s = \alpha_s KTV_I - (\delta_s + \mu_s)T_s,$$

$$\frac{d}{dt}T_l = \alpha_l KTV_I - (\delta_l + \mu_l)T_l,$$

$$\frac{d}{dt}T_p = \alpha_p KTV_I + \delta_l T_l - \delta_p T_p, \tag{18.2}$$

$$\frac{d}{dt}V_I = (1 - \eta)(P + \delta_m T_m + \delta_s T_s + \delta_p T_p) - eV_I,$$

$$\frac{d}{dt}V_{NI} = \eta(P + \delta_m T_m + \delta_s T_s + \delta_p T_p) - cV_{NI}.$$

In order to make the units in the various compartments comparable, we have expressed everything in terms of *viron equivalents*. Thus, the number of T_m cells, say, is N, the number of virions produced during its lifetime times the number of such T cells. This is incorporated into the constant infection rate K.

Note that system (18.2) is a linear compartmental system if all the rates including T are assumed to be constant. Moreover, note that $\delta_m + \mu_m$, $\delta_s + \mu_s$, and $\delta_l + \mu_l$ can just be lumped into flow rates, σ_m, σ_s, and σ_l, respectively. Likewise, ηP and $(1-\eta)P$ can be lumped into forcing function terms f_1 and f_2.

Using further biological information, Wu and Ding proceed to simplify system (18.2) and eventually obtain a closed-form solution to the modified system; that is, the equation corresponding to V_I is reduced to

$$\frac{d}{dt}V_I = (1 - \eta)P - cV_I.$$

Note that this equation can be solved analytically since it simply corresponds to a nonhomogeneous first-order ordinary differential equation. Having solved for V_I, this is then plugged into equations in terms of T_m, T_s, T_l, and T_p, which are now solvable in closed form. Once these solutions are obtained, these are plugged into the last remaining equation, $\frac{d}{dt}V_{NI}$, and this is likewise solved.

In the presence of treatment by antiviral drugs, protease inhibitors and reverse transcriptase inhibitors, the dynamics of the infection change. The equation involving the infectious virus is reduced to

$$\frac{d}{dt}V_I = (1 - \eta)P - cV_I$$

since the production from infected cells is small compared to the constant production P. The other equations are modified by reducing the flow from the V_I compartment by a factor $(1-\gamma)$ to account for the reduced infection rate.

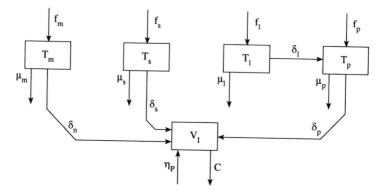

Figure 18.5: The model of HIV infection in the presence of treatment. The inputs f_m, f_s, f_l, and f_p involve the quantity V_I and are of the form $f_m = (1-\gamma)\alpha_m kTV_I$, etc.

The V_I equation can be solved analytically and substituted in each of the T equations to find a closed-form solution for the system. The compartmental model is shown in Figure 18.5. The V_I compartment is no longer needed since it can be considered a forcing function for the T compartments. The inspected T cells which are also part of this forcing function are assumed to follow an exponential recovery function $T = a_1 + (a_0 - a_1)e^{-rt}$.

A simpler model involving only one compartment for infected cells was proposed by Perelson et al. (1997). Its compartmental model is given in Figure 18.6. Again, the V_I compartment does not depends on the others

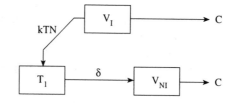

Figure 18.6: A simple model of HIV infection.

and the level may be treated as an input to a two-compartment system. We also again adopt the convention that the quantity of infected T-cells are in units of virions produced during each cell's lifetime.

Problem 18.4

Write down the differential equations for the simple model of Figure 18.6. Assuming the number of uninfected T cells is given by an expression $T = a_1 + (a_0 - a_1)e^{-rt}$, find a solution to this system.

Part IV

Compartmental Models Theory

In the previous chapters, we have been concerned chiefly with particular examples and special cases of compartmental models. However, many of the properties of the special cases carry over to the general case as well. This is particularly true for the linear models. We shall see that the structure of the digraph to some extent determines properties of the solution of the differential equations.

19

Basic Properties of Linear Models

As we saw in Chapter 13, the equation of a compartmental model with donor-controlled flow rates has the form

$$\frac{d\mathbf{x}}{dt} = A\mathbf{x} + \mathbf{f}_i - E_o\mathbf{x}, \tag{19.1}$$

whereas that with recipient-controlled flow rates has the form

$$\frac{d\mathbf{x}}{dt} = B\mathbf{x} + E_i\mathbf{x} - \mathbf{f}_o. \tag{19.2}$$

The matrices A and B are both singular since the sum of the columns is zero in both cases. The difference is that all diagonal elements in A are nonpositive while the off-diagonal elements are non-negative, whereas in B, the opposite is true. The matrices E_o and E_i are both diagonal matrices with non-negative elements.

A particular compartmental model may have a mixture of donor- and recipient-controlled flow rates. This is true in an ecosystem model, in which an introduced predator may have it easy going for a while, i.e., be recipient controlled, when the rest of the system is living with scarcity, i.e., donor controlled. With tracer experiments models, the flow rates are only donor controlled since a fraction of the donor compartments passes to the recipient compartments.

We shall try to emulate the simple two-compartment case considered in Chapter 13, and shall see that the results there are general. For donor-controlled systems, the solution is non-negative and converges to a stable equilibrium solution.

19.1   Compartmental Matrices

By a compartmental matrix, we strictly mean a matrix of the form of the sum of the matrices in (19.1), i.e., of the form

$$C = A - E_o. \tag{19.3}$$

Thus, C will have non-negative off diagonal and nonpositive diagonal elements and each column sum will be ≤ 0. The matrix corresponding to $E_o = 0$ yields a matrix C, which is singular as previously discussed. Moreover, while B and $B + E_i$ are not compartmental matrices, their negatives are, so theoretical results will extend to recipient-controlled models as well.

The matrix function

$$Z(t) = e^{Ct}$$

is used to solve the differential equation (19.1). It may be defined in a number of ways (see Appendix A.2). The easiest is by means of the convergent infinite series of matrices

$$e^{Ct} = I + Ct + \frac{C^2 t^2}{2!} + \cdots + \frac{C^k t^k}{k!} + \cdots. \tag{19.4}$$

This series shares many of the properties of the scalar exponential summation. In particular, we have, if A and B commute,

$$e^{C \cdot 0} = e^O = I \quad \text{and} \quad e^{At} e^{Bt} = e^{(A+B)t}.$$

However, it is not always the case that $e^{Ct} \geq 0$ as it would be in the scalar case; nonetheless, in Proposition 19.1 we will show that this is the case when C is a compartmental matrix.

In order to do numerical calculations with e^{Ct}, it will usually be of the form obtained from (19.4) by means of eigenvalues and eigenvectors of the matrix C.

We assume that C is an $n \times n$ matrix with a complete set of eigenvectors $\mathbf{k}_1, \mathbf{k}_2, \ldots, \mathbf{k}_n$ with associated eigenvalues $\lambda_1, \lambda_2, \ldots, \lambda_n$, which for simplicity of illustration we assume to be distinct; thus,

$$C\mathbf{k}_i = \lambda_i \mathbf{k}_i, \quad i = 1, 2, \ldots, n.$$

This may be expressed in matrix form as

$$CK = K\Lambda, \tag{19.5}$$

where $K = [\mathbf{k}_i, \mathbf{k}_2, \ldots, \mathbf{k}_n]$ is the matrix whose columns are the eigenvectors and $\Lambda = \text{diag}\{\lambda_1, \lambda_2, \ldots, \lambda_n\}$. Then, since our eigenvalues are all distinct, it follows that K is of full rank so that we may form the spectral decomposition, that is,

$$C = K\Lambda K^{-1}$$

and, hence,

$$C^2 = (K\Lambda K^{-1})(K\Lambda K^{-1}) = K\Lambda^2 K^{-1}.$$

Similarly,

$$C^k = (K\Lambda^k K^{-1}) = K \operatorname{diag}\{\lambda_1^k, \lambda_2^k, \dots, \lambda_n^k\} K^{-1}. \qquad (19.6)$$

We now substitute this into (19.4), and by employing the fact that Λ is a diagonal matrix, we obtain

$$e^{Ct} = I + K\Lambda t K^{-1} + \frac{K\Lambda^2 t^2}{2!} K^{-1} + \cdots + \frac{K\Lambda^k t^k}{k!} K^{-1} + \cdots$$

$$= K \left(I + \Lambda t + \frac{\Lambda^2 t^2}{2!} + \cdots + \frac{\Lambda^k t^k}{k!} + \cdots \right) K^{-1}$$

$$= K \operatorname{diag} \left(\sum_{k=0}^{\infty} \frac{\lambda_1^k t^k}{k!}, \dots, \sum_{k=0}^{\infty} \frac{\lambda_n^k t^k}{k!} \right) K^{-1}$$

$$= K e^{\Lambda t} K^{-1}. \qquad (19.7)$$

As an example, we return to the 2×2 matrix

$$A = \begin{bmatrix} -a & b \\ a & -b \end{bmatrix}$$

with eigenvalues 0 and $-(a+b)$. We find the corresponding eigenvectors that form the columns of K; that is,

$$K = \begin{bmatrix} b & 1 \\ a & -1 \end{bmatrix},$$

with corresponding inverse matrix

$$K^{-1} = \frac{1}{b+a} \begin{bmatrix} 1 & 1 \\ a & -b \end{bmatrix}$$

and substituting this into (19.7), we have

$$e^{At} = \frac{1}{b+a} \begin{bmatrix} b & 1 \\ a & -1 \end{bmatrix} \begin{bmatrix} 1 & 0 \\ 0 & e^{-(b+a)t} \end{bmatrix} \begin{bmatrix} 1 & 1 \\ a & -b \end{bmatrix}$$

$$= \frac{1}{b+a} \begin{bmatrix} b + ae^{-(b+a)t} & b - be^{-(b+a)t} \\ a - ae^{-(b+a)t} & a + be^{-(b+a)t} \end{bmatrix}.$$

which is exactly the expression which appeared in our solution in Chapter 13. Note that in the above 2×2 case $e^{At} > 0$, since b, a, and $e^{-at} > 0$ while $e^{-at} \le 1$; furthermore, this is true, in general, for compartmental matrices.

PROPOSITION 19.1
Let C be a compartmental matrix. Then $e^{Ct} \geq 0$ for all $t > 0$.

PROOF   Let $h > 0$ be a number such that $C + hI$ has positive diagonal elements. Then $C + hI$ has all elements ≥ 0 and hence

$$e^{(C+hI)t} = \sum_{k=0}^{\infty} (C + hI)^k \frac{t^k}{k!} \geq 0.$$

But from this and since $h > 0$ it follows that

$$e^{(C)t} = e^{(C+hI)t} e^{-hIt} = e^{(C+hI)t} I e^{-ht} \geq 0,$$

since $(C + hI)$ and I commute. Hence our proposition is established.   ∎

Problems 19.1

1. Show that the solution to the differential equation (19.1) is strictly positive for all t when $\mathbf{f}_i = \mathbf{0}$, $E_0 = 0$,

$$A = \begin{bmatrix} -1 & 1 \\ 1 & -1 \end{bmatrix},$$

   and $\mathbf{x}(0) > \mathbf{0}$.

2. For $B = -A$, A as in Problem 1, show that for $\mathbf{f}_0 = \mathbf{0}$ and $E_i = 0$ in (19.2), there is a solution to the differential equation which has a negative component whenever $\mathbf{x}(0) \neq \mathbf{0}$.

19.2   Eigenvalues

In our previous 2×2 case, we had two real eigenvalues 0 and $-(b + a)$ and both were real and nonpositive. For the most general 2×2 case, the compartmental matrix can be seen to be

$$C = A - E_o = \begin{bmatrix} -a - c & b \\ a & -b - d \end{bmatrix},$$

where a, b, c, and d are all non-negative. The eigenvalues are now both real, distinct, and nonpositive.

In general, things can get more complex as the number of compartments increases. The three-compartment system of Figure 19.1 has as its matrix

$$A = \begin{bmatrix} -\alpha & 0 & \alpha \\ \alpha & -\alpha & 0 \\ 0 & \alpha & -\alpha \end{bmatrix}.$$

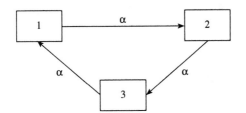

Figure 19.1: A closed three-compartment system.

Its eigenvalues are easily found and they can be seen to satisfy the equation $(\lambda + \alpha)^3 - \alpha^3 = \lambda(\lambda^2 + 3\lambda\alpha + 3\alpha^2) = 0$. Thus, they are

$$\lambda_1 = 0, \quad \lambda_{2,3} = \alpha \left(-\frac{3}{2} \pm \frac{\sqrt{3}}{2} i \right).$$

Hence, we cannot conclude that the eigenvalues of a compartmental matrix are necessarily real. Again, in this case, the eigenvalues are distinct and have a nonpositive real part. The following proposition tells us that this result is general.

PROPOSITION 19.2
Let C be the compartmental matrix of a model with donor (respectively, recipient-)-controlled flows. Then, all the nonzero eigenvalues of C have a negative (respectively, positive) real part.

PROOF   The location of the eigenvalues of a matrix may be estimated by Gersgorin's theorem from matrix theory (which can be found in Marcus and Minc (1964), p. 446). It says that each of the eigenvalues of a matrix are contained in the union of all disks in the complex plane of the form

$$|\lambda - c_{jj}| \le \sum_{i=1}^{n} |c_{ij}|, \quad j = 1, \ldots, n, \ i \ne j. \tag{19.8}$$

In the donor-controlled case, $c_{jj} \le 0$, $j = 1, 2, \ldots, n$. If $c_{jj} = 0$, then all flows leaving compartment j are zero as well, and the disk (19.8) reduces to a single point $\lambda = 0$. If $c_{jj} < 0$, each of the disks has the form

$$|\lambda + \gamma| \le \alpha,$$

where $\gamma > 0$ and $\alpha \le \gamma$, and hence lies entirely in the half-plane $\lambda \le 0$. The only possible eigenvalues with a zero real part are zero itself since $|i\beta + \gamma| \le \gamma$ implies $\beta = 0$.

In the recipient-controlled case, all the signs are reversed and all eigen-values must lie in disks of the form $|\lambda - \beta| \leq \delta$, where $\delta \leq \beta$. Thus, our proposition is established. However, we point out that in the recipient-controlled case, the corresponding matrix is not a compartmental matrix as defined in Section 19.1. ∎

Problems 19.2

1. Show by example that a closed three-compartment system as in Figure 19.1 but with different flow rates can have real eigenvalues.

2. Show that for a three-compartment system whose digraph has a catenary form, the eigenvalues are real.

19.3   Analytic Solution

We now return to Equations (19.1) and (19.2) and consider the properties of the solutions to the differential equations. We first consider the homogeneous versions in which $\mathbf{f}_c$ (or $\mathbf{f}_o$) are zero. Then, the equations become

$$\frac{d\mathbf{x}}{dt} = C\mathbf{x},$$

$$\mathbf{x}(0) = \mathbf{x}_o, \tag{19.9}$$

where C is a compartmental matrix or its negative. Again, we assume that C has a complete set of eigenvectors so that (19.5) holds. We then introduce a change of variable

$$\mathbf{x} = K\mathbf{y}, \qquad \mathbf{x}_o = K\mathbf{y}_o,$$

which converts (19.9) into

$$K\frac{d\mathbf{y}}{dt} = CK\mathbf{y} = K\Lambda\mathbf{y}$$

or

$$\frac{d\mathbf{y}}{dt} = \Lambda\mathbf{y}. \tag{19.10}$$

This is a decoupled system and has the solution

$$\mathbf{y} = e^{\Lambda t}\mathbf{y}_o,$$

where $e^{\Lambda t}$ is as in (19.7). This may be transformed back into $\mathbf{x}$ to get

$$\mathbf{x} = Ke^{\Lambda t}K^{-1}\mathbf{x}_o = e^{Ct}\mathbf{x}_o \tag{19.11}$$

as our solution. However, it is more instructive to express this in the form

$$\mathbf{x} = c_1 \mathbf{k}_1 e^{\lambda_1 t} + \cdots + c_n \mathbf{k}_n e^{\lambda_n t}, \tag{19.12}$$

where $c_1, c_2, \ldots, c_n$ are constants which by (19.12) for $t = 0$ must satisfy

$$\mathbf{x}_o = K \begin{bmatrix} c_1 \\ c_2 \\ \vdots \\ c_n \end{bmatrix}.$$

In the donor-controlled case, if one of the eigenvalues, say, $\lambda_1 = 0$, then $\mathbf{x}(t) \to c_1 \mathbf{k}_1$ as $t \to \infty$. (In the recipient-controlled case $\mathbf{x}(t)$ will be unbounded.) From Proposition 20.1, we will see that if $\mathbf{x}_o \geq \mathbf{0}$, then $\mathbf{x}(t) \geq \mathbf{0}$ for all $t \geq 0$. Thus, even though $\mathbf{x}(t)$ may be oscillatory, none of the components becomes negative.

If the model (donor controlled) has a differential equation of the form

$$\frac{d\mathbf{x}}{dt} = A\mathbf{x} - E\mathbf{x} + \mathbf{f}, \tag{19.13}$$

where $\mathbf{f}$ is the vector of flows in, then the solution may be found by the method of variation of parameters or through integrating factors as follows. As usual, we can express the solution to (19.13) as the sum of a particular solution $\mathbf{x}_p$ and the general solution to the homogeneous equation (19.9). The particular solution is found by replacing the initial constants in (19.9) by variables, i.e., setting

$$\mathbf{x} = e^{Ct}\mathbf{u}.$$

This is then substituted into (19.13) with $C = A - E$ to obtain

$$Ce^{Ct}\mathbf{u} + e^{Ct}\frac{d\mathbf{u}}{dt} = Ce^{Ct}\mathbf{u} + \mathbf{f}$$

since the product rule for differentiation extends to matrix products. Then, $\frac{d\mathbf{u}}{dt}$ satisfies

$$\frac{d\mathbf{u}}{dt} = e^{-Ct}\mathbf{f}$$

and a particular solution is

$$\mathbf{x}_p(t) = e^{Ct} \int_0^t e^{-Cs}\mathbf{f}(s)\,ds.$$

As an example, consider the model given by the diagram in Figure 19.2. The corresponding differential equation is

$$\mathbf{x}' = \begin{bmatrix} -1 & 1 \\ 1 & -1 \end{bmatrix}\mathbf{x} - \begin{bmatrix} 0 & 0 \\ 0 & 2 \end{bmatrix}\mathbf{x} + \begin{bmatrix} 1 \\ 0 \end{bmatrix}$$

$$= \begin{bmatrix} -1 & 1 \\ 1 & -3 \end{bmatrix}\mathbf{x} + \begin{bmatrix} 1 \\ 0 \end{bmatrix}.$$

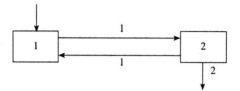

Figure 19.2: A simple model with nonzero input.

Then, a particular solution is

$$\mathbf{x}_p(t) = e^{Ct} \int_0^t e^{-Cs} \begin{bmatrix} 1 \\ 0 \end{bmatrix} ds$$

$$= e^{Ct}(-C^{-1})(e^{-Ct} - I) \begin{bmatrix} 1 \\ 0 \end{bmatrix}$$

$$= e^{Ct}e^{-Ct}(-C^{-1}) \begin{bmatrix} 1 \\ 0 \end{bmatrix} + C^{-1}e^{Ct} \begin{bmatrix} 1 \\ 0 \end{bmatrix}$$

since e^{-Ct} commutes with C^{-1}. Hence, since the second term is a solution to the homogeneous equation, we have

$$\mathbf{x}_p = \frac{-1}{2} \begin{bmatrix} -3 & -1 \\ -1 & -1 \end{bmatrix} \begin{bmatrix} 1 \\ 0 \end{bmatrix} = \begin{bmatrix} \frac{3}{2} \\ \frac{1}{2} \end{bmatrix},$$

which can be checked by substitution.

Problem 19.3

Modify the last example by taking the flows out of compartment 2 to be 0 ($E = 0$). The matrix C will then be singular. Find a particular solution in this case.

20

Structure and Dynamical Properties

In the last chapter, we saw that compartmental matrices had eigenvalues whose real part, if not zero, was negative. Hence, any solution to the associated differential equation was asymptotically stable. We also saw that positive initial values lead to non-negative solutions. However, there is still the possibility that a solution have a zero component in finite time. We also would like to determine structural conditions under which the asymptotic levels are nonzero; i.e. conditions for which $\mathbf{x}(t) > \mathbf{0}$ as $t \to \infty$. We also consider a method for simplifying the structure. Certain cases with a particular structure are shown to have associated dynamical behavior.

20.1 Positivity of Solutions

Again, we suppose that

$$\mathbf{x}' = C\mathbf{x} + \mathbf{f}, \quad \mathbf{x}(0) = \mathbf{x}_o \geq \mathbf{0}, \quad \mathbf{f} \geq \mathbf{0}, \tag{20.1}$$

where C is a compartmental matrix. We already know that $\mathbf{x} \geq \mathbf{0}$, but, in fact, we have the following stronger result.

PROPOSITION 20.1
Let $\mathbf{x}$ satisfy (20.1); then, for each component of $\mathbf{x}$, either $x_i(t) = 0$ for all $t \geq 0$ or there is some $t_i \geq 0$ such that $x_i(t) > 0$ for all $t > t_i$, and $x_i(t) = 0$ for $t < t_i$. In particular, if $x_i(0) > 0$, then $x_i(t) > 0$ for all $t > 0$.

PROOF Either $x_i(t) = 0$, or there is some time τ such that $x_i(\tau) > 0$. In the latter case, form the function $x_i(t)\phi_i(t)$, where

$$\phi_i(t) = e^{-c_{ii}t}.$$

Then, from the definition of $\frac{dx_i}{dt}$ and the above, we see that the product rule yields

$$\frac{d(x_i\phi_i)}{dt} = \sum_{j=1}^{n}(c_{ji}x_j - c_{ii}x_i)\phi_i + f_i$$

$$= \sum_{j=1, j\neq i}^{n} c_{ji}x_j + f_i \geq 0, \qquad (20.2)$$

since $c_{ji} > 0$ for $i \neq j$ and $x_i\phi_i$ is monotonically nondecreasing. Hence, $x_i(t)\phi_i(t) \geq x_i(\tau)\phi_i(\tau)$ for $t > \tau$. Since $\phi_i(t) > 0$, it follows that $x_i(t) > 0$ whenever $x_i(\tau) > 0$. Let $t_i = \inf(\tau \mid x_i(\tau) > 0)$. Then, $x_i(t) = 0$ for $t < t_i$ since $x_i(t) \geq 0$ for all t. This completes the proof. ∎

This proposition shows us that a particular compartment may never get any of the material if it started that way, i.e., if $x_i(0) = 0$. This could happen if the compartment has flows out but none in. We would expect, however, that a strongly connected digraph would allow the introduction of material to each compartment, and this is, in fact, the case.

PROPOSITION 20.2
Let $\mathbf{x}$ satisfy (20.1), let the digraph D of the compartmental model be strongly connected; then, if $\mathbf{x}(0) \neq \mathbf{0}$, then $\mathbf{x}(t) > \mathbf{0}$ for all $t > 0$.

PROOF Since $\mathbf{x}(0) \neq \mathbf{0}$, there is at least one compartment such that $x_i(0) > 0$. By Proposition 20.1, $x_i(t) > 0$ for all $t > 0$. Let j be any other compartment; then, by the strong connectivity, there is a path from i to j passing through vertices $u_1, u_2, \ldots, u_k$. Again, by using the same trick as in (20.2), we find that

$$\frac{dx_{u_1}(t)\phi_{u_1}(t)}{dt} \geq a_{iu_1}(t)\phi_{u_1}(t) > 0$$

for $t > 0$. Hence, $x_{u_1}(t) > 0$ for $t > 0$. We repeat the argument again to show $x_{u_2}(t) > 0, x_{u_3}(t) > 0, \ldots, x_{u_j}(t) > 0$. Thus, our proposition is established. ∎

Although the strong connectivity is sufficient for the positivity of all components, it is possible to come up with weaker conditions. We use a procedure similar to the previous proof to get the following result.

COROLLARY 20.3
Let $\mathbf{x}$ satisfy (20.1) and let $x_i(0) > 0$ for some compartment i; if all compartments (vertices) are reachable from i, then $\mathbf{x}(t) > \mathbf{0}$ for all $t > 0$.

In (20.1), we have not required that the input be constant, but it has been in all our examples. The following is an example in which the input is not constant. Let

$$\mathbf{x}' = \begin{bmatrix} -1 & 1 \\ 1 & -1 \end{bmatrix} \mathbf{x} + \begin{bmatrix} e^{-t} \\ 0 \end{bmatrix}.$$

In this case, the eigenvalues–eigenvector decomposition of e^{Ct} is

$$e^{\begin{bmatrix} -1 & 1 \\ 1 & -1 \end{bmatrix} t} = \frac{1}{2} \begin{bmatrix} 1 & 1 \\ 1 & -1 \end{bmatrix} \begin{bmatrix} 1 & 0 \\ 0 & e^{-2t} \end{bmatrix} \begin{bmatrix} 1 & 1 \\ 1 & -1 \end{bmatrix}^{-1}$$

$$= \frac{1}{2} \begin{bmatrix} 1 + e^{-2t} & 1 - e^{-2t} \\ 1 - e^{-2t} & 1 + e^{-2t} \end{bmatrix}.$$

The particular solution $\mathbf{x}_p$ is given by

$$\mathbf{x}_p = e^{Ct} \int_0^t e^{-Cs} \begin{bmatrix} e^{-s} \\ 0 \end{bmatrix} ds$$

$$= e^{Ct} \int_0^t e^{-(C+I)s} \begin{bmatrix} 1 \\ 0 \end{bmatrix} ds$$

$$= e^{Ct} \left[e^{-(C+I)t} - e^{-(C+I)0} \right] (-C - I)^{-1} \begin{bmatrix} 1 \\ 0 \end{bmatrix}$$

$$= \begin{bmatrix} \frac{1 - e^{-2t}}{2} \\ \frac{(1 - e^{-t})^2}{2} \end{bmatrix}.$$

In this section, we have seen that a structural concept, strong connectivity of the digraph, leads to a positive stable equilibrium solution to the differential equation. If the digraph is not strongly connected, we look at the various strongly connected components to try to understand the dynamics.

Problem 20.1

Show that for the model of Figure 19.1, any nontrivial, non-negative solution is positive.

20.2   Condensation of the Digraph

In Chapter 3, we studied the strong components of a digraph and introduced the concept of condensation of a digraph. In this section, we extend the definition to compartmental models. We assume these are weakly connected and are closed, i.e., have no flows from or to the outside. This is really no restriction as far as flows to the outside are concerned in the donor-controlled case. We could simply add another compartment labeled outside and have a closed system. The input can sometimes also be handled similarly if it consists of a "bolus" input, that is, an input impulse at time $t = 0$. This can be effectively translated to the initial condition, as we will see later. Therefore, our result is of the form

$$\frac{d\mathbf{x}}{dt} = A\mathbf{x}, \quad \mathbf{x}(0) = \mathbf{x}_o, \tag{20.3}$$

where A is such that $[\,1\quad 1\quad 1\quad \cdots\quad 1\,]A = [\,0\quad 0\quad 0\quad \cdots\quad 0\,]$, i.e., the columns of A add up to zero. This happens because the diagonal elements of A are just $-\sum_{j=1, i \neq j}^{n} a_{ij}$, as we have seen previously. Thus, A is singular and must have at least one eigenvalue equal to zero.

Let D be the digraph of the model and let D^* be the condensation digraph whose vertices are the strong components of D. We can associate with D^* a compartmental model whose flow rates are just the average of those leaving each component; that is, the flow rate a_{ij}^* from component K_i to K_j is given by

$$a_{ij}^* = \frac{\sum a(K_i, K_j)}{\#(K_i)}, \tag{20.4}$$

where the sum is taken over all arcs (K_i, K_j) from K_i to K_j and $\#(K_i)$ is just the number of compartments in K_i.

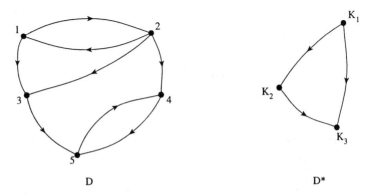

D   D*

Figure 20.1: An example digraph and its condensations.

In the example of Figure 20.1, the compartmental matrix is of the form

$$A = \begin{bmatrix} a_{11} & a_{21} & 0 & 0 & 0 \\ a_{12} & a_{22} & 0 & 0 & 0 \\ a_{13} & a_{23} & a_{33} & 0 & 0 \\ 0 & a_{24} & 0 & a_{44} & a_{54} \\ 0 & 0 & a_{35} & a_{45} & a_{55} \end{bmatrix}.$$

The compartmental matrix of D^* is

$$A^* = \begin{bmatrix} a_{13}^* - a_{12}^* & 0 & 0 \\ a_{12}^* & -a_{23}^* & 0 \\ a_{13}^* & a_{23}^* & 0 \end{bmatrix}.$$

where

$$a_{12}^* = \frac{a_{13} + a_{31}}{2},$$

$$a_{13}^* = \frac{a_{24}}{2},$$

$$a_{23}^* = a_{35}.$$

In this example, the only possible ordering is K_1, K_2, K_3. A more complex example is given in Figure 20.2. The ordering there could be K_1, K_5, K_2, K_3, K_6, K_7, K_4, K_8, but this is not unique; another ordering would be K_5, K_1, K_2, K_6, K_3, K_7, K_4, K_8. In both cases, there are no arcs going from later vertices to earlier ones.

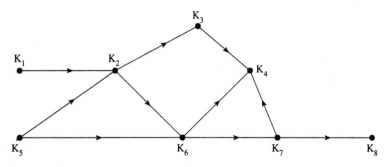

Figure 20.2: An acyclic condensation digraph.

The strong components ordered in this way lead to a partition of the original matrix A. In the first example, we have

$$A = \begin{bmatrix} a_{11} & a_{21} & 0 & 0 & 0 \\ a_{12} & a_{22} & 0 & 0 & 0 \\ \hline a_{13} & a_{23} & a_{33} & 0 & 0 \\ \hline 0 & a_{24} & 0 & a_{44} & a_{54} \\ 0 & 0 & a_{35} & a_{45} & a_{55} \end{bmatrix}$$

$$= \begin{bmatrix} A_{11} & O & O \\ A_{12} & A_{22} & O \\ A_{13} & A_{23} & A_{33} \end{bmatrix}.$$

This always leads to such a block triangular matrix. The eigenvalues of such matrices are found by calculating the eigenvalues of the individual digraph block matrices, $A_{11}, A_{22}, A_{33}, \ldots$. This follows from the fact that the determinant of A is

$$\det A = \det A_{11} \det A_{22} \cdots.$$

PROPOSITION 20.4
λ *is an eigenvalue of* A *if and only if it is an eigenvalue of a strongly connected component of* A.

PROOF   The conclusion follows from the partitioning of $A - \Lambda I$ and applying the above expression to $\det(A - \Lambda I)$.   ∎

As we saw in Chapter 12, we can convert a closed compartmental model to a Markov chain by approximating the derivative with a difference quotient. This enables us to use results from Markov chain theory for our compartmental models. Then, (20.3) can be written

$$\mathbf{x}(t + h) \approx \mathbf{x}(t) + h\,A\mathbf{x}(t)$$

or

$$\mathbf{x}^T(t + h) \approx \mathbf{x}^T(t)(I + hA)^T. \tag{20.5}$$

Then, if h is sufficiently small, the columns of $I + hA$ add up to 1 and the elements are non-negative. Hence,

$$P = (I + hA)^T$$

is a stochastic matrix. It may be partitioned by the strongly connected components just as A was, and the condensation may be converted into a stochastic matrix as well.

Let P^* be this matrix; it must have the form

$$P^* = \begin{bmatrix} Q & R \\ O & I \end{bmatrix}$$

since the vertices in the vertex contrabasis have no arcs leaving them. In the two examples, Figures 20.1 and 20.2, the vertex contrabasis is K_3 in the first and $\{K_4, K_8\}$ in the second.

Clearly, P^* is the matrix of an absorbing Markov chain. Hence, any material is ultimately absorbed in the absorbing states. Each of these absorbing states is a strong component in a digraph. To understand them, we shall look at the behaviors of models with strongly connected digraphs in the next section.

20.3 Eigenvalues and Structure

For a closed (donor-controlled) compartmental model, that is, one with no flows from or to the outside, the compartmental matrix is always singular. Hence, it has zero as an eigenvalue. If its digraph is strongly connected, then it has multiplicity 1. All the remaining $(n-1)$ eigenvalues have a negative real part but may be complex, as in Figure 19.1. In this section, we consider the spread of the nonzero eigenvalues. This is important for identifiability (in the next chapter) and for questions related to the stability of ecosystems.

We only consider matrices of strongly connected compartmental models (SCCM). We shall use a number of results from matrix theory which may be found in the text by Marcus and Minc (1964).

DEFINITION 20.1   *The* sum of squares deviation(SS) *of the nonzero eigenvalues of the compartmental matrix A of an SCCM is the number*

$$SS = \sum_{i=2}^{n} \left| \lambda_i - \frac{1}{n-1} \sum_{i=2}^{n} \lambda_i \right|^2,$$

where $\lambda_2, \lambda_3, \ldots, \lambda_n$ *are the nonzero eigenvalues of A.*

PROPOSITION 20.5
Let A be an SCCM matrix. Then,

$$\left| \mathrm{Tr}\, A^2 - \frac{(\mathrm{Tr}\, A)^2}{n-1} \right| \leq SS \leq \mathrm{Tr}(AA^T) - \frac{(\mathrm{Tr}\, A)^2}{n-1}$$

with equality on the left if the eigenvalues are real.

Here, $\mathrm{Tr}\, A$ *is the trace* $\mathrm{Tr}\, A = \sum a_{ii}$ *and* A^T *the transpose of A.*

PROOF   Since the trace is invariant under similarity transformation, we have

$$\operatorname{Tr} A = \sum \lambda_i, \quad \operatorname{Tr} A^2 = \sum \lambda_i^2, \qquad (20.6)$$

and, therefore,

$$\left| \operatorname{Tr} A^2 - \frac{(\operatorname{Tr} A)^2}{n-1} \right| = \left| \sum \lambda_i^2 - \frac{\left(\sum \lambda_i \right)^2}{n-1} \right|$$

$$= \left| \sum \lambda_i^2 - (n-1)\bar{\lambda}^2 \right| = \left| \sum (\lambda_i - \bar{\lambda})^2 \right|$$

$$\leq \sum |\lambda_i - \bar{\lambda}|^2,$$

where $\bar{\lambda}$ is the mean value of the nonzero λ_i's. By Schur's inequality (Marcus and Minc (1964), p. 142),

$$\sum |\lambda_i|^2 \leq \sum a_{ij}^2 = \operatorname{Tr}(AA^T),$$

and since $\bar{\lambda}$ is real and therefore

$$\sum |\lambda_i - \bar{\lambda}|^2 = \sum |\lambda_i|^2 - (n-1)(\bar{\lambda})^2,$$

the conclusion follows.   ∎

REMARK 20.6   The inequalities are the best possible, since the inequality on the right is attained for a SCCM whose digraph is a cycle of length 3. Indeed, its matrix is

$$A = \begin{bmatrix} -a & 0 & c \\ a & -b & 0 \\ 0 & b & -c \end{bmatrix},$$

and for $a = b = c = 1$, the nonzero eigenvalues are

$$\lambda = \frac{-3 \pm \sqrt{3}i}{2}.$$

Hence, $|\lambda_1|^2 + |\lambda_2|^2 = 2(\frac{9}{4} + \frac{3}{4}) = 6$ and $\operatorname{Tr}(AA^T) = 2(a^2 + b^2 + c^2) = 6$.

There are a number of other matrices associated with an SCCM matrix A. One is the adjoint matrix A^* composed of cofactors. Another is the matrix B, a nonsingular $(n-1) \times (n-1)$ matrix with the same nonzero eigenvalues as A.

PROPOSITION 20.7
Let A^* be the adjoint matrix of the SCCM matrix A. Then, each column sum of A^* is the same as

$$\text{Tr}\, A^* = \prod_{i=2}^{n} \lambda_i.$$

PROOF Since the rank of A is $n-1$, the rank of A^* must be 1, and since $AA^* = O = A^*A$, it follows that each row of A^* is a multiple of $[1\ 1\cdots 1]$. (Recall that $[1\ 1\cdots 1]A = [0\ 0\cdots 0]$.)

Let $\mathbf{w}$ be a normalized equilibrium solution of (19.1) (with $\mathbf{f}_i = \mathbf{0}$ and $E_0 = O$), i.e., $A\mathbf{w} = \mathbf{0}$ and $[1\ 1\cdots 1]\mathbf{w} = 1$. Then, each column of A^* is the same multiple of $\mathbf{w}$, and A^* has the form

$$A^* = \alpha[\mathbf{w}\ \mathbf{w}\cdots \mathbf{w}],$$

whence $\alpha = \text{Tr}\, A^*$.

In the characteristic polynomial $p(\lambda)$ of A, the coefficient of $-\lambda$ is given by $\prod_{i=2}^{n} \lambda_i$ (since $\lambda_1 = 0$). Since it is also given by $\text{Tr}\, A^*$, the conclusion follows. ∎

The other matrix associated with A is obtained by a simple operation.

DEFINITION 20.2 *Let B be the $(n-1) \times (n-1)$ matrix obtained by subtracting the first column of A from every other column and then removing the first row and column.*

PROPOSITION 20.8
If A is the SSCM matrix, then B, as given in Definition 20.2, is a nonsingular matrix whose nonzero eigenvalues agree with those of A.

PROOF Let $\lambda \neq 0$, $\mathbf{x}$ be a characteristic pair of A. We partition both A and $\mathbf{x}$ so that

$$\lambda \mathbf{x} = \lambda \begin{bmatrix} x_1 \\ \mathbf{x}_2 \end{bmatrix} = \begin{bmatrix} A_{11} & A_{12} \\ A_{21} & A_{22} \end{bmatrix} \begin{bmatrix} x_1 \\ \mathbf{x}_2 \end{bmatrix},$$

where A_{11} is a 1×1 submatrix and x_1 is the first component of $\mathbf{x}$. Then,

$$\lambda \mathbf{x}_2 = A_{21} x_1 + A_{22} \mathbf{x}_2,$$

and since $x_1 = -\sum_{i=2}^{n} x_i = -[1, \ldots, 1]\mathbf{x}_2$, it follows that

$$\lambda \mathbf{x}_2 = (A_{22} - [1, \ldots, 1]A_{21})\mathbf{x}_2 = B\mathbf{x}_2.$$

∎

This may be used in conjunction with standard theorems on the spread of a matrix (Marcus and Minc (1964) p. 167) to obtain a bound on the spread on the nonzero eigenvalues of A.

COROLLARY 20.9
Let $\lambda_2, \ldots, \lambda_n$ be the nonzero eigenvalues of A. Then,

$$\max_{i,j} |\lambda_i - \lambda_j| \leq \left(2\|B\|^2 - \frac{2}{n-1}(\mathrm{Tr}\, A)^2 \right)^{1/2}.$$

Here, the norm of B is the l^2 norm,

$$\|B\|^2 = \sum_{i,j=1}^{n-1} b_{ij}^2.$$

Example
Let A be given by

$$A = \begin{bmatrix} -4 & 3 & 1 & 1 \\ 1 & -3 & 1 & 2 \\ 2 & 0 & -2 & 1 \\ 1 & 0 & 0 & -4 \end{bmatrix}.$$

Then, B will be

$$B = \begin{bmatrix} -4 & 0 & 1 \\ -2 & -4 & -1 \\ -1 & -1 & -5 \end{bmatrix}$$

and

$$\max_{i,j} |\lambda_i - \lambda_j| \leq \left[2 \times 65 - \tfrac{1}{2}(13)^2 \right]^{1/2}$$

$$= \left(130 - 112\tfrac{1}{3} \right)^{1/2} = \left(17\tfrac{2}{3} \right)^{1/2} = 4.2.$$

This should be contrasted with the spread of the eigenvalues of A (including the 0 eigenvalue), which would be about 6.9. The sum of the eigenvalues is -13, and their product is $\mathrm{Tr}\, A^* = (-24 - 31 - 27 - 6) = -78$.

20.4   Some Special Cases

We study the eigenvalues of special cases of SCCMs in which the digraph is that of a catenary system or a mammillary system. In each case, we decompose the sum of squares deviation into the deviation of the eigenvalues of the individual cycles plus other terms arising from connections.

In a catenary system, the digraph has the form given in Figure 20.3, in which each vertex is connected by arcs in both directions to the previous and subsequent vertex and no others. Its matrix is in triple diagonal form with positive entries directly above and below the main diagonal and negative entries on it. Hence, all eigenvalues are real and simple (Marcus and Minc (1964), p. 166). Hence, by Proposition 20.5, the sum of squares deviation of the eigenvalues is given by

$$SS = \text{Tr } A^2 - \left(\frac{\text{Tr } A}{n-1} \right)^2 .$$

The matrix A is given by

$$A = \begin{bmatrix} -a_1 & b_1 & 0 & \cdots & 0 \\ a_1 & -a_2 - b_3 & b_2 & \cdots & 0 \\ 0 & a_2 & -a_3 - b_3 & \cdots & 0 \\ \vdots & \vdots & \vdots & & \vdots \\ 0 & 0 & 0 & \cdots & b_{n-1} \\ 0 & 0 & 0 & \cdots & -b_{n-1} \end{bmatrix} .$$

The trace of A^2 may be simplified to get

$$\text{Tr } A^2 = \sum_{i=1}^{n-1} (a_i + b_i)^2 + 2 \sum_{i=1}^{n-2} a_{i+1} b_i,$$

whereas

$$(\text{Tr } A)^2 = \left(\sum_{i=1}^{n-1} a_i + b_i \right)^2 .$$

Hence,

Figure 20.3: Digraph of a catenary system with five compartments.

$$SS = 2 \sum_{i=1}^{n-2} a_{i+1} b_i + \sum_{i=1}^{n-1} (a_i + b_i)^2 - \frac{1}{n-1} \left(\sum_{i=1}^{n-1} (a_i + b_i) \right)^2 . \tag{20.7}$$

This equation may be interpreted in terms of the eigenvalues of the matrices of the individual cycles. Each cycle has a matrix of the form

$$
A_i =
\begin{bmatrix}
O & \big| & O & \big| & O \\
\hline
O & \begin{matrix} -a_i & b_i \\ a_i & -b_i \end{matrix} & O \\
\hline
O & \big| & O & \big| & O
\end{bmatrix},
$$

all of whose eigenvalues are zero except

$$
\lambda^{(i)} = -(a_i + b_i) = \operatorname{Tr} A_i.
$$

There are $n-1$ of these cycles. Hence, the mean value of all their eigenvalues is

$$
\frac{1}{n-1} \sum_{i=1}^{n-2} \lambda^{(i)} = \frac{1}{n-1} \sum_{i=1}^{n-1} -(a_i + b_i) = \frac{1}{n-1} \operatorname{Tr} A = \bar{\lambda},
$$

which is the same as the mean value of the nonzero eigenvalues of A. Hence, (20.7) becomes

$$
\begin{aligned}
\operatorname{SS} &= 2 \sum_{i=1}^{n-2} a_{i+1} b_i + \sum_{i=1}^{n-1} (\lambda^{(i)} - \bar{\lambda})^2 \\
&= \operatorname{SS}_c + \operatorname{SS}_b;
\end{aligned}
\tag{20.8}
$$

that is, the total sum of squares deviation is equal to the sum of squares due to connections of the cycles (SS_c) plus the sum of the squares between the cycles (SS_b). The latter is, of course, zero if the eigenvalues of the cycles are all the same.

The product of the eigenvalues may, by Proposition 20.7, be obtained from the trace of A^*. This, in turn, may be found by first finding an equilibrium solution $\mathbf{w}$ and then rescaling it so that the first component is the first cofactor of A. The result is that the product is given by

$$
\prod_{i=1}^{n} \lambda_i = \operatorname{Tr} A^* = \sum_{i=1}^{n} a_i a_2 \cdots a_{i-1} b_i \cdots b_{n-1}.
\tag{20.9}
$$

The compartmental model for a mammillary system has a central vertex which lies on all cycles, each of which is of length 2. (See Figure 20.4 for an example.) Its matrix has the form

$$
A =
\begin{bmatrix}
-a_{11} & b_1 & \cdots & b_{n-1} \\
a_1 & -b_1 & \cdots & O \\
O_2 & O & \cdots & O \\
\vdots & \vdots & & \vdots \\
a_{n-1} & O & \cdots & -b_{n-1}
\end{bmatrix},
$$

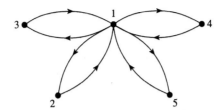

Figure 20.4: A mammillary system with five vertices.

where $a_{11} = -a_1 - a_2 - \cdots - a_{n-1}$. Its trace is given by

$$\operatorname{Tr} A = -\sum_{i=1}^{n-1}(a_i + b_i),$$

whereas

$$\operatorname{Tr} A^2 = \sum_{i=1}^{n-1}(a_i + b_i)^2 + 2\sum_{i=1}^{n-2} a_i \sum_{j=i+1}^{n-1} a_j.$$

Again, it may be shown that the eigenvalues are real and, hence, we can get a closed-form expression for SS. The lower bound from Proposition 20.5 is

$$\text{SS} = 2\sum_{\substack{i=1 \\ i<j}}^{n-2} a_i a_j + \sum_{i=1}^{n-1}(a_i + b_i)^2 - \frac{1}{n-1}\left(\sum_{i=1}^{n-1} a_i + b_i\right)^2. \qquad (20.10)$$

Just as with the catenary system, the digraph may be subdivided into individual cycles of length 2, whose individual nonzero eigenvalues are again denoted by $\lambda^{(i)}$. Then, we have, since $\lambda^{(i)} = -(a_i + b_i)$,

$$\text{SS} = \sum_{\substack{i=1 \\ i<j}}^{n-2} a_i a_j + \sum_{i=1}^{n-1}(\lambda^{(i)} - \bar{\lambda})^2$$

$$= \text{SS}_c + \text{SS}_b. \qquad (20.11)$$

In both this special case and the previous one, the SS_c term is a sum of terms, each of which is the product of two flow rates leaving the same compartments.

The product of the eigenvalues may again be calculated from the trace of the adjoint A^*. In this case, it is

$$\prod_{i=1}^{n-1}\lambda_i = \prod_{i=1}^{n-1}(-b_i)\left(1 + \frac{a_1}{b_1} + \frac{a_2}{b_2} + \cdots + \frac{a_{n-1}}{b_{n-1}}\right). \qquad (20.12)$$

We shall use both types of results in Chapter 23, where we consider stability and complexity indices.

Problems 20.4

1. A catenary system with three compartments is also a mammillary system. Show that (20.7) and (20.8) give the same values in this case.

2. Compare the deviation of the eigenvalues for a catenary system and a mammillary system with constant flow rates, $b_i = a_i = 1$, $i = 1, 2, \ldots, n - 1$.

3. Find the eigenvalues for the two cases in Problem 2 when $n = 4$.

21

Identifiability of a Compartmental System

In most of the work in the previous chapters, we have assumed a knowledge of the flow rates and have tried to find properties of the solution to the system of differential equations. In this chapter, we attack the inverse problem: Given a solution, what are the properties of the flow rates?

21.1 General Input and Output

The input to a compartmental system is usually to a single compartment. However, it may, in certain circumstances, be mixed before it is distributed to input compartments. This involves an input matrix which we denote by B (not to be confused with the matrix used in recipient-controlled models or in the previous chapter). For example, in a two-compartment system, the matrix $B = [1 \ 1]$ would be used to indicate that the input is shared equally by compartments 1 and 2.

In general, we assume that the input to the system is $\mathbf{f} = B\mathbf{u}$. Typically, B has one or two columns but has as many rows as there are compartments.

The output similarly may be combined. However, since our system is assumed donor controlled, the output will have the form

$$\mathbf{y} = C\mathbf{x},$$

in which the output of the compartments is directed to one or more measuring devices with measurements $\mathbf{y}(t)$. These outputs are observations rather than flows out of the system. Here, C is now a general sampling matrix of such observations, which is different from the previous matrices with the same name. The compartmental matrix is now denoted by A.

The structure of the model will be as in Figure 21.1.

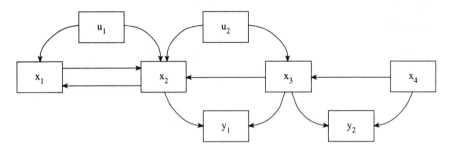

Figure 21.1: A four-compartment system with two inputs and two outputs.

The problem then is translated to the differential equation form

$$\frac{d\mathbf{x}}{dt} = A\mathbf{x} + B\mathbf{u},$$

$$\mathbf{y} = C\mathbf{x},$$

$$\mathbf{x}(0) = \mathbf{0}, \qquad\qquad (21.1)$$

where we have assumed the initial levels are 0. This latter is really no restriction for a "bolus" input which can be treated either as an initial value or incorporated into $B\mathbf{u}$.

The properties of the equation are determined by the triple (A, B, C) of matrices. The techniques discussed in previous chapters enables us to find a solution of (21.1) if all the matrices are known. It is

$$\mathbf{y}(t) = C\mathbf{x}(t) = C \int_0^t e^{A(t-s)} B\mathbf{u}(s)ds,$$

where $\mathbf{x}(0) = \mathbf{0}$. Typically, however, the flow rates in the matrix A are unknown and must be determined from a knowledge of $\mathbf{u}(t)$ and $\mathbf{y}(t)$. The problem of determining such an A is known as the *identification problem*. A system is identifiable if there is a unique solution to this problem. However, it does not always have a solution and even when it does, it may be difficult to find. The problem of actually estimating the parameters of an identifiable problem is an ill-posed problem; that is, the output may be close to the true output, whereas the parameters are still quite different than the true parameters; that is, such parameters are usually sensitive to slight perturbations of the data. This is analogous with the problems encountered with the linear system $A\mathbf{x} = \mathbf{b}$, where one wishes to solve for $\mathbf{x}$ from knowledge of $\mathbf{b}$ and A, and A is ill-conditioned.

21.2 An Example of an Identifiable System

We return to the simple two compartment example given by the diagram in Figure 21.2.

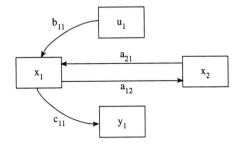

Figure 21.2: A simple two-compartment system.

The triple (A, B, C) is

$$A = \begin{bmatrix} -a_{12} & a_{21} \\ a_{12} & -a_{21} \end{bmatrix}, \quad B = \begin{bmatrix} b_{11} \\ 0 \end{bmatrix}, \quad C = [c_{11} \quad 0].$$

We take $b_{11} = c_{11} = 1$ and ask if a_{12} and a_{21} are identifiable. Rather than take a general input, we assume that

$$u(t) = \delta(t - \epsilon),$$

a known bolus input immediately (ϵ) after the system is started. Such *delta functions* or "impulses" are given approximately by functions of the form

$$\delta_\lambda(t) = \lambda h(\lambda t)$$

as $\lambda \to \infty$, where $h(t) = 1$ if $0 < t \leq 1$, otherwise $h(t) = 0$. The limit is taken in the weak sense; that is,

$$\lim_{\lambda \to \infty} \int_{-\infty}^{\infty} \delta_\lambda(t)\phi(t)dt = \int_{-\infty}^{\infty} \delta(t)\phi(t)dt = \phi(0),$$

where $\phi(t)$ is continuous near zero. These delta functions have many interesting properties (see Zemanian (1965)), one of which is the fact that

$$\int_{-\infty}^{\infty} \delta(t - s)\phi(s)ds = \phi(t)$$

i.e., δ is the identity of convolution. Returning now to our system, we see that

$$y(t) = C \int_0^t e^{A(t-s)} B\delta(s - \epsilon)ds$$

$$= Ce^{A(t-\epsilon)}B.$$

Then, by taking the limit as $\epsilon \to 0$, we get the *impulse response*

$$y(t) = Ce^{At}B$$

$$= \begin{bmatrix} 1 & 0 \end{bmatrix} K \begin{bmatrix} e^{\lambda_1 t} & 0 \\ 0 & e^{\lambda_2 t} \end{bmatrix} K^{-1} \begin{bmatrix} 1 \\ 0 \end{bmatrix}, \qquad (21.2)$$

where λ_1 and λ_2 are the eigenvalues of A, and K is the eigenvector matrix. The eigenvalues are $\lambda_1 = 0$ and $\lambda_2 = -a_{12} - a_{21}$ in this case, and K is the matrix

$$K = \begin{bmatrix} a_{21} & 1 \\ a_{12} & -1 \end{bmatrix}, \quad K^{-1} = \begin{bmatrix} -1 & 1 \\ -a_{12} & a_{21} \end{bmatrix} \frac{1}{(-a_{21} - a_{12})}.$$

Thus, we have

$$y(t) = \begin{bmatrix} a_{21} & 1 \end{bmatrix} \begin{bmatrix} 1 & 0 \\ 0 & e^{-(a_{21}+a_{12})t} \end{bmatrix} \begin{bmatrix} -1 \\ a_{21} \end{bmatrix} \frac{1}{(a_{21}+a_{12})}$$

$$= \begin{bmatrix} a_{21} & e^{\lambda_2 t} \end{bmatrix} \begin{bmatrix} 1 \\ a_{12} \end{bmatrix} \frac{1}{a_{21}+a_{12}}$$

$$= \frac{a_{21} + a_{12}e^{\lambda_2 t}}{a_{21} + a_{12}}.$$

This shows our system to be identifiable. We can find $\frac{a_{21}}{a_{21}+a_{12}}$ from very large values of t since $e^{\lambda_2 t} \to 0$ as $t \to \infty$. Then, any other value of t gives us a second equation, which combined with the first, gives us a unique solution

$$a_{21} = (a_{21} + a_{12})y(\infty),$$

$$a_{21} + a_{12}e^{\lambda_2(t_1)} = (a_{21} + a_{12})y(t_1).$$

Note, however, that the second equation is really

$$a_{21} + a_{12}e^{-(a_{12}+a_{21})(t_1)} = (a_{21} + a_{12})y(t_1),$$

i.e. is *transcendental* since it involves a_{12} and a_{21} in the exponent.

21.3   Another Example Which Is Not Identifiable

We now consider the same example and hypothesize that both compartments have flows to the outside. The system of equations has matrices

$$A = \begin{bmatrix} -a_{12} - a_{10} & a_{21} \\ a_{12} & -a_{21} - a_{20} \end{bmatrix}, \quad B = \begin{bmatrix} 1 \\ 0 \end{bmatrix}, \quad C = \begin{bmatrix} 1 & 0 \end{bmatrix}.$$

Now however, the eigenvalues of A are both negative; moreover, there are four parameters to estimate. Another way to obtain equations in these parameters is to differentiate (21.2) repeatedly. Then, we get $y(0) = 1$, and

$$y'(0) = CAB,$$
$$y''(0) = CA^2B,$$
$$y'''(0) = CA^3B,$$

$$\vdots \quad \vdots$$

The first two equations we have are

$$-(a_{12} + a_{10}) = y'(0)$$

$$(a_{12} + a_{10})^2 - a_{21}a_{12} = y''(0),$$

but then become more complicated. Fortunately, we can obtain a third independent equation by integration instead of differentiation:

$$\int_0^\infty y(t)dt = \int_0^\infty Ce^{At}Bdt$$
$$= Ce^{At} A^{-1}B\big|_0^\infty$$
$$= -CA^{-1}B;$$

it is still insufficient for us to estimate four parameters. In fact, additional equations will be redundant as we shall see in the next section, which involves an alternate approach.

21.4 Another Approach: Using Laplace Transforms

The Laplace transform, defined for functions of at most exponential growth, is given by

$$L(f(t)) = \int_0^\infty e^{-st}f(t)dt := F(s).$$

See, e.g., Bellman et al. (1966), for properties. The derivative satisfies

$$L(f'(t)) = sF(s) - f(0),$$

as can be shown by integration by parts. For the examples of the last section,

$$y(t) = Ce^{At}B,$$

and therefore

$$Y(s) = C \int_0^\infty e^{At} e^{-sIt} dt B$$
$$= C(-(A - sI)^{-1})B$$
$$= C(sI - A)^{-1}B.$$

In these particular cases, we can express $Y(s)$ as

$$Y(s) = \frac{Q(s)}{P(s)},$$

which we find, in the first example, to be

$$Y(s) = C \begin{bmatrix} s + a_{12} & a_{21} \\ -a_{12} & s + a_{22} \end{bmatrix}^{-1} B$$

$$= [1 \quad 0] \frac{1}{(s + a_{12})(s + a_{21}) - a_{12}a_{21}}$$

$$\times \begin{bmatrix} s + a_{12} & a_{21} \\ a_{12} & s + a_{22} \end{bmatrix} \begin{bmatrix} 1 \\ 0 \end{bmatrix}$$

$$= \frac{s + a_{21}}{s^2 + (a_{12} + a_{21})s}$$

$$= \frac{s + \alpha_1}{s^2 + \beta_1 s + \beta_2},$$

where $\alpha_1 = a_{21}$, $\beta_2 = 0$ and $\beta_1 = a_{12} + a_{21}$. Thus, from this $Y(s)$, we can determine a_{12} and a_{21}.

For the second example, we get

$$\frac{Q(s)}{P(s)} = \frac{s + a_{21} + a_{20}}{s^2 + (a_{21} + a_{20} + a_{12} + a_{10})s + (a_{21}a_{10} + a_{20}a_{12} + a_{20}a_{10})}$$

and three equations in four unknowns:

$$a_{12} + a_{20} = \alpha_1,$$

$$a_{21} + a_{20} + a_{12} + a_{10} = \beta_1,$$

$$a_{21}a_{10} + a_{20}a_{12} + a_{20}a_{10} = \beta_2.$$

In this case, we again have no unique solution to the problem and the system is not identifiable.

We might ask if there is any way in which this model is identifiable. A plausible approach might be to add another measurement y_2 at compartment 2. Then, the matrix C becomes a 2×2 identity matrix and

$$Y(s) = C(sI - A)^{-1}B$$

a 2×1 matrix. After a standard calculation, we see that

$$\begin{bmatrix} Y_1(s) \\ Y_2(s) \end{bmatrix} = \frac{\begin{bmatrix} s + a_{21} + a_{20} \\ -a_{21} \end{bmatrix}}{s^2 + (a_{12} + a_{21} + a_{20} + a_{02})s + a_{21}a_{10} + a_{12}a_{20} + a_{20}a_{10}}$$

Thus, from the numerator of $Y_1(s)$, we can find $a_{21} + a_{20}$, and from that of $Y_2(s)$, we can find a_{21}. The denominator gives us two additional equations which can be used to find a_{10} and a_{12} thereby making our new system identifiable.

Problem 21.3

Show by the method of Laplace transforms that the example in Section 21.2 is identifiable.

21.5   The General Case

In a general model with n compartments, the associated compartmental matrix A is $n \times n$. If the model has p inputs and q outputs, then B will be a $n \times p$ matrix and C a $q \times n$ matrix. The observed measurements $y(t)$ constitute a $q \times p$ matrix, again given by

$$y(t) = Ce^{At}B.$$

Its Laplace transform is

$$Y(s) = C(sI - A)^{-1}B.$$

The inverse $(sI - A)^{-1}$ may be shown, by expanding in minors (see Marcus and Minc (1964)), to be of the form

$$(sI - A)^{-1} = \frac{Q(s)}{P(s)}$$

where $P(s)$ is the characteristic polynomial of A and $Q(s)$ is a polynomial of degree at most $(n-1)$ with $q \times p$ matrix coefficients. Thus, $Y(s)$ has the same form

$$Y(s) = \frac{1}{P(s)}CQ(s)B, \tag{21.3}$$

except the numerator has $q \times p$ matrix coefficients.

If $Y(s)$ is known, then (21.3) gives us n equations in the coefficients in A from the denominator alone. The numerator gives us $n - 1$ equations for each element in the matrix. The total number of equations is therefore

$n + pq(n - 1)$. The total number of unknown flows may be as large as n^2 if each compartment has flows to every other and to the outside. This will be the case if nothing at all is known about the structure of the digraph. This gives the negative result.

PROPOSITION 21.1

If the total number of nonzero flows in the compartmental model exceeds $n + pq(n - 1)$, then the system is not identifiable: In particular, if the structure is not specified, the system is not identifiable if $n > pq$.

The last inequality holds if $n^2 > n + pq(n - 1)$.

Sufficient conditions are usually quite technical since even if the number of equations is greater than or equal to the number of flows, there may be redundancy among the equations. This happens in the case of a cycle which has exactly $n - 1$ flows (provided it has no excretions) but leads to $2n - 2$ equations. For necessary and sufficient conditions, see, for example, Cobelli and Romanin-Jacur (1975), Delforge (1981), Walter (1982), Anderson (1983), and Walter (1987).

21.6 Size Identifiability of Compartmental Models

In most approaches to the study of identifiability of the parameters of a compartmental model, the size is assumed known. Yet, in many biomedical applications, it is unclear whether to include certain compartments in the model. Of course, a lower bound on the number of compartments is given by the number of distinct exponential decay terms observed. However, this is clearly not an upper bound if the model has a compartment which is not output reachable, i.e., has no path to a compartment with an output. In this case, any number of compartments which are not output reachable may be added without influencing the output. If, on the other hand, the digraph of the model is strongly connected, then it is not obvious whether it is possible to have more compartments than distinct exponential decays (when the eigenvalues are distinct).

This then is the problem with which we shall be concerned: When can a compartmental model with a strongly connected digraph of size n have fewer than n exponential decays? We assume that the eigenvalues are distinct, since if they are not, the answer is obvious. The exponential decays refers to distinct nonzero terms in the impulse response $\phi(t) = Ce^{At}B$.

Unfortunately, it is very easy to construct counterexamples even in the case of distinct eigenvalues, and we shall mention a few of them; that is, it is possible to find A together with input and output matrices B and C

such that $\phi(t) = \sum_{k=1}^{n} R_k e^{\lambda_k t}$ has at least one zero R_k.

However, these counterexamples are all quite pathological in that they only work when there is some functional relation among the transfer coefficients. Therefore, we shall modify our question to ask whether it is *probable* that $\phi(t)$ have fewer than n distinct nonzero terms for a given structure. If a probability measure is put on the matrices A that is consistent with Euclidean measure, then we shall be interested in cases where the probability is one that $\phi(t)$ have n distinct terms. In probability terminology, such statements are true "almost surely."

For a compartmental matrix A, the concept of *irreducibility* and strong connectivity are equivalent. The matrix is *reducible* if by a permutation of its rows and columns it can be put into the form

$$A = \begin{bmatrix} A_{11} & O \\ A_{21} & A_{22} \end{bmatrix}.$$

The associated digraph is clearly not strongly connected since there are no flows from the vertices associated with A_{22} to those associated with A_{11}. The converse is also true, since by using the condensation of the digraph as we did in Chapter 20, we can show that a digraph which is not strongly connected has an associated matrix of this form.

We shall assume initially that A has distinct eigenvalues $\lambda_1, \lambda_2, \ldots, \lambda_n$ with right eigenvectors $\mathbf{p}_1, \mathbf{p}_2, \ldots, \mathbf{p}_n$ and left eigenvectors $\mathbf{s}_1^T, \mathbf{s}_2^T, \ldots, \mathbf{s}_n^T$. This assumption will be justified by results to come. We shall assume that they are normalized such that $\mathbf{s}_i^T \mathbf{p}_i = 1$, $i = 1, 2, \ldots, n$. The matrix of right and left eigenvectors will be denoted by

$$P = [\mathbf{p}_1, \mathbf{p}_2, \ldots, \mathbf{p}_n], \quad S = [\mathbf{s}_1, \mathbf{s}_2, \ldots, \mathbf{s}_n]^T.$$

Then, clearly, we have

$$AP = P\Lambda,$$
$$SA = \Lambda S,$$
$$SP = I,$$

where Λ is the diagonal matrix of eigenvalues.

We also have the *spectral resolution* of A,

$$f(A) = \sum_{k=1}^{n} f(\lambda_k) Z_k, \tag{21.4}$$

where Z_k is a matrix whose columns are right eigenvectors and whose rows are left eigenvectors corresponding to the eigenvalue λ_k. It is given by

$$Z_k = \mathbf{p}_k \mathbf{s}_k^T. \tag{21.5}$$

We have the following simple proposition on reducibility of P and S.

PROPOSITION 21.2
If A is irreducible, so are P and S.

PROOF If P is reducible of the form

$$P = \begin{bmatrix} P_{11} & O \\ P_{21} & P_{22} \end{bmatrix}.$$

then P_{11} and P_{22} are nonsingular and, hence,

$$P^{-1} = \begin{bmatrix} P_{11}^{-1} & O \\ Q_{21} & P_{22}^{-1} \end{bmatrix}$$

is also reducible of the same form. Since the main diagonal is contained in the two matrices P_{11} and P_{22}, it follows that Λ has the same form. Hence,

$$A = P\Lambda P^{-1}$$

has the same form as well and is reducible. The same argument holds for S as well. ∎

The conclusion of this proposition does not hold for the Z_k. Some of them may well be reducible when A is not. Consider, for example, the following simple system.

21.6.1   Example

Let A be the matrix of the model given by

$$A = \begin{bmatrix} -1 & 1 & 1 \\ 1 & -2 & 1 \\ 0 & 1 & -2 \end{bmatrix}.$$

Its eigenvalues are $\lambda = 0, -2, -3$, its right eigenvectors are

$$\mathbf{p}_1 = \begin{bmatrix} 3 \\ 2 \\ 1 \end{bmatrix}, \quad \mathbf{p}_2 = \begin{bmatrix} 1 \\ 0 \\ -1 \end{bmatrix}, \quad \mathbf{p}_3 = \begin{bmatrix} 0 \\ 1 \\ -1 \end{bmatrix},$$

and its left eigenvectors are

$$\mathbf{s}_1^T = \tfrac{1}{6}[1 \quad 1 \quad 1], \quad \mathbf{s}_2^T = \tfrac{1}{2}[1 \quad -1 \quad -1], \quad \mathbf{s}_3^T = \tfrac{1}{3}[-1 \quad 2 \quad -1].$$

Hence, we find

$$Z_1 = \mathbf{p}_1 \mathbf{s}_1^T = \frac{1}{6}\begin{bmatrix} 3 & 3 & 3 \\ 2 & 2 & 2 \\ 1 & 1 & 1 \end{bmatrix},$$

$$Z_2 = \frac{1}{2} \begin{bmatrix} 1 & -1 & -1 \\ 0 & 0 & 0 \\ -1 & 1 & 1 \end{bmatrix}, \quad Z_3 = \frac{1}{3} \begin{bmatrix} 0 & 0 & 0 \\ -1 & 2 & -1 \\ 1 & -2 & 1 \end{bmatrix}.$$

Clearly, both Z_2 and Z_3 are reducible, while A is not.

This example also furnishes a counterexample to the original question. If $B^T = [1 \quad 0 \quad 0]$, then

$$e^{At}B = \frac{1}{6} \begin{bmatrix} 3 \\ 2 \\ 1 \end{bmatrix} + \frac{e^{-2t}}{2} \begin{bmatrix} 1 \\ 0 \\ -1 \end{bmatrix} + \frac{e^{-3t}}{3} \begin{bmatrix} 0 \\ -1 \\ 1 \end{bmatrix}.$$

Hence, if $C = [1 \quad 0 \quad 0]$, then $Ce^{At}B = \frac{1}{2}(1 + e^{-2t})$, whereas if $C = [0 \quad 1 \quad 0]$, we have $Ce^{At}B = \frac{1}{3}(1 - e^{-3t})$.

21.6.2 Some More Examples

The simplest nontrivial case of a compartmental model with strongly connected digraph is one with a 2×2 matrix A of the form

$$A = \begin{bmatrix} -a_{11} & a_{12} \\ a_{21} & -a_{22} \end{bmatrix}, \quad a_{ij} > 0, \ i,j = 1,2, \quad a_{11} \geq a_{21}, \ a_{22} \geq a_{12}.$$

In this case, the results are obvious. Let $B^T = [1 \quad 0]$ and $C = [1 \quad 0]$ or $[0 \quad 1]$; then, the impulse response ϕ is

$$\phi(t) = Ce^{At}B = r_1 e^{\lambda_1 t} + r_2 e^{\lambda_2 t},$$

where $r_1 r_2 \neq 0$ and $\lambda_1 \neq \lambda_2$.

In the general case, the eigenvalues are no longer necessarily distinct, nor is the conclusion true for every single input and output. However, we do have a simple result for certain B and C when there are n distinct eigenvalues.

PROPOSITION 21.3

Let A be any $n \times n$ compartmental matrix with distinct eigenvalues; then, B (C) is a column (row) vector not orthogonal to any left (right) eigenvector of A if and only if

$$\phi(t) = \sum_{i=1}^{n} r_i e^{\lambda_i t},$$

where

$$\prod_{i=1}^{n} r_i \neq 0.$$

PROOF   We use the components of A to write

$$Ce^{At}B = C\sum_{k=1}^{n}e^{\lambda_k t}Z_k B = \sum_{k=1}^{n}C\mathbf{p}_k\mathbf{s}_k^T Be^{\lambda_k t}.$$

If $C\mathbf{p}_k \neq 0$ and $\mathbf{s}_k^T B \neq 0$, then $r_k \neq 0$, and conversely. Hence, the conclusion follows.   ∎

COROLLARY 21.4
Let A be as in Proposition 21.3; then the conclusion holds for almost all row vectors C and column vectors B.

PROOF   The set of row vectors orthogonal to at least one right eigenvector of A is a set of measure zero in the set of all $C \geq 0$; similarly for B.   ∎

We should like to strengthen Proposition 21.3 to allow single input and output matrices C. However, the following example shows we cannot do so.

Example
Let A be the matrix of the catenary system (or mammillary system) given by

$$A = \begin{bmatrix} -1 & 1 & 0 \\ 1 & -2 & 1 \\ 0 & 1 & -1 \end{bmatrix}.$$

Then, A has eigenvalues $\lambda = 0, -1, -3$, right eigenvector matrix

$$P = \begin{bmatrix} 1 & 1 & 1 \\ 1 & 0 & -2 \\ 1 & -1 & 1 \end{bmatrix},$$

and left eigenvector matrix

$$S = \begin{bmatrix} \frac{1}{3} & \frac{1}{3} & \frac{1}{3} \\ \frac{1}{2} & 0 & -\frac{1}{2} \\ \frac{1}{6} & -\frac{1}{3} & \frac{1}{6} \end{bmatrix}.$$

Thus, e^{At} is given by

$$e^{At} = \frac{1}{3}\begin{bmatrix} 1 & 1 & 1 \\ 1 & 1 & 1 \\ 1 & 1 & 1 \end{bmatrix} + \frac{e^{-t}}{2}\begin{bmatrix} 1 & 0 & -1 \\ 0 & 0 & 0 \\ -1 & 0 & 1 \end{bmatrix} + \frac{e^{-3t}}{6}\begin{bmatrix} 1 & -2 & 1 \\ -2 & 4 & -2 \\ 1 & -2 & 1 \end{bmatrix}.$$

Hence, if $B^T = \begin{bmatrix} 0 & 1 & 0 \end{bmatrix}$, then for all C, $\phi(t) = Ce^{At}B$ contains only two nonzero terms. Also, if $C = \begin{bmatrix} 0 & 1 & 0 \end{bmatrix}$, then for all B, $\phi(t)$ contains only two nonzero terms as well. If either B^T or C is taken to be $[1 \quad 1 \quad 1]$, then $\phi(t)$ is just a constant. Of course, in both of these cases, either B or C is orthogonal to some of the eigenvectors.

21.6.3 Main Result

We now address the question raised at the start. Among all models with a given size and structure (i.e., digraph D), what is the portion whose impulse response contains fewer than n distinct decay terms? We shall assume that there is a single bolus input and a single output, i.e., that B and C are coordinate vectors, so that $\phi(t)$ is a scalar.

REMARK 21.5 Since, in the distinct eigenvalue case,

$$\phi(t) = Ce^{At}B$$
$$= \sum_{k=1}^{n} e^{\lambda_k t} CZ_k B, \qquad (21.6)$$

it will contain fewer than n terms if $CZ_k B = 0$ for some Z_k. Since both C and B are coordinate vectors, $CZ_k B$ is merely an element of Z_k. But $Z_k = \mathbf{p}_k \mathbf{s}_k^T$, where $\mathbf{s}_k^T$ is a left eigenvector and $\mathbf{p}_k$ is a right eigenvector. Hence, either $\mathbf{s}_k$ or $\mathbf{p}_k$ contains a zero component if Z_k does, and, in fact, Z_k will have either a row or a column of zeros (or both). Hence, we shall look for eigenvectors with zero components.

In the second example, Z_2 has both a row and column of zeros, whereas in the first example, it has a zero row but not a zero column.

The principal eigenvalue, i.e., the one with smallest modulus, always has positive left and right eigenvectors when D is strongly connected (see Marcus and Minc (1964)). Hence, the term corresponding to it will always appear in $\phi(t)$. This is not necessarily true for other eigenvalues, as the examples show.

We now return to the general case and let A be an $n \times n$ irreducible compartmental matrix with a given digraph D but with arbitrary positive coefficients. The set of all such matrices is isomorphic to R_+^N, where N is the number of arcs including excretions in D. A property will be said to hold for "almost all" A if it holds for all of R_+^N except for a set of Euclidean measure 0.

Our major result is the following.

THEOREM 21.6
Let D be the digraph of a compartmental model of size n, which is strongly connected; then, for almost all compartmental matrices A with digraph D

and for all single input and output matrices B and C, the impulse response

$$\phi(t) = Ce^{At}B$$

contains n distinct exponential terms.

The proof will be omitted. It may be found in Walter (1986) and involves showing that almost all of the eigenvectors have no zeros.

22

Parameter Estimation

Identifiability, as discussed in the last chapter, is primarily a theoretical concept. It may be possible in theory to find the flow rates given the output but may be impossible in practice. There are a number of reasons for this. One is that the output is usually only known at discrete times and, then, only approximately. This must be used to estimate the true output or its Laplace transform. Another reason is that the equations involving the coefficients must be solved numerically since they are coupled nonlinear equations. Still another reason is that the exact structure of the model is seldom known so that the equations to be estimated may not be correct.

22.1 Estimation Problem

We discuss three approaches to estimation. We begin with the assumption that we are given a discrete set of observations/measurements made at times t_i, $i = 1, \ldots, m$, such that

$$y_i = \phi(t_i) + \epsilon_i, \tag{22.1}$$

where the error ϵ_i is assumed to be normally distributed with true mean 0 and some variance, $\sigma \geq 0$, which we assume to be constant and known (more is said about variance estimation in a later section). In (22.1), we are assuming that the only unknown in $\phi(t)$ is the compartmental matrix A; hence, the sampling and input matrices, C and B, respectively, are known. Here, $\phi(t)$ is the theoretical response solving (21.1).

The standard approach to estimation is to solve for A in the *least squares* sense in (22.1); that is, to find $A = [a_{kj}]$ which gives a solution to

$$\min_{a_{kj}} \sum_i |y_i - \phi(t_i)|^2 \tag{22.2}$$

subject to the condition that A is a compartmental matrix of the appropriate model. See Bates and Watts (1988) and Audoly, D'Angio, Saccomani, and Cobelli (1998) and the references found there in, for instance.

Another approach to estimation involves two stages (Anderson (1983)). In this, the data are first fitted to a sum of exponentials

$$g(t) = \sum_{k=1}^{n} c_k e^{\lambda_k t}.$$

By the last section of Chapter 21, the number of exponentials to be considered is the same as the size of the model. The estimates $\hat{c}_k$ and $\hat{\lambda}_k$ are obtained by least squares as in (22.2), by finding a solution to

$$\min_{c_k, \lambda_k} \sum_i |y_i(t_i) - g(t_i)|^2, \quad k = 1, 2, \ldots, n, \tag{22.3}$$

subject to $\lambda_k \leq 0$. Then, the Laplace transform of the fitted function $g(t)$ is found; it has the form

$$\hat{G}(s) = \sum_{k=1}^{n} \frac{\hat{c}_k}{s - \hat{\lambda}_k}. \tag{22.4}$$

This, in turn, is compared to the Laplace transform of $\phi(t)$ as in Section 21.4,

$$\Phi(s) = C(sI - A)^{-1} B. \tag{22.5}$$

Since both $\hat{G}(s)$ and $\Phi(s)$ are rational functions, both the numerator and denominator are polynomials whose coefficients must be equal. To illustrate this, we return to the example in Section 21.3.

This example has two compartments with an input to the first compartment. It is identifiable when there is one excretion from either compartment 1 or 2 but not both. We assume, therefore, that $a_{20} = 0$; then,

$$A = \begin{bmatrix} -a_{12} - a_{10} & a_{21} \\ a_{12} & -a_{21} \end{bmatrix}$$

and $B^T = [1\ 0] = C$ as before and $u(t) = \delta(t)$. This means that the input into compartment 1 is a one-time bolus input, and the sampling is from that same compartment; so our impulse response function is then

$$\Phi(s) = C(sI - A)^{-1} B = \frac{s + \gamma_1}{s^2 + \gamma_2 s + \gamma_3}, \tag{22.6}$$

where

$$\gamma_1 = a_{12}, \quad \gamma_2 = a_{21} + a_{10} + a_{12}, \quad \gamma_3 = a_{10} a_{12}.$$

Notice that the directed graph of A is strongly connected so that, by the results established in the previous section, we can say with probability 1 that $\phi(t)$ contains two distinct exponential terms.

Thus, we first fit the data to an arbitrary sum-of-exponentials involving two terms; that is, we solve the following problem:

$$\min_{c_1, c_2, \lambda_1, \lambda_2} \sum_{i=1}^{m} (y_i - g(t_i))^2,$$

subject to $\lambda_1 \leq 0$, $\lambda_2 \leq 0$, to get

$$\hat{g}(t) = \hat{c}_1 e^{\hat{\lambda}_1 t} + \hat{c}_2 e^{\hat{\lambda}_2 t}. \tag{22.7}$$

Then,

$$\hat{G}(s) = \frac{\hat{c}_1}{s - \hat{\lambda}_1} + \frac{\hat{c}_2}{s - \hat{\lambda}_2} = \frac{s + \hat{\beta}_1}{s^2 + \hat{\beta}_2 s + \hat{\beta}_3}. \tag{22.8}$$

Since (22.6) and (22.8) must be equal, and two rational expressions are equal if and only if their coefficients are equal, we get that

$$\gamma_1 = \hat{\beta}_1, \quad \gamma_2 = \hat{\beta}_2, \quad \gamma_3 = \hat{\beta}_3, \tag{22.9}$$

where

$$\hat{\beta}_1 = -(\hat{c}_1 \hat{\lambda}_2 + \hat{c}_2 \hat{\lambda}_1), \quad \hat{\beta}_2 = -(\hat{\lambda}_1 + \hat{\lambda}_2), \quad \hat{\beta}_3 = \hat{\lambda}_1 \hat{\lambda}_2.$$

We see that (22.9) yields a nonlinear system of equations in terms of the unknowns a_{12}, a_{21}, and a_{10}. In this case, a closed-form solution can be found, but, in general, a scheme such as Newton's method is needed. The reader is referred to Dennis and Schnabel (1983) or Björck (1996) for details of numerical methods involving Newton updates.

There is also the method of *peeling*, which avoids least squares. This involves fitting an exponential function to the later data values. This corresponds to the eigenvalue of smallest magnitude. The difference between the data and this function is then used to approximate the second eigenvalue, etc. (see Jacquez (1996)).

Still another method involves direct Laplace transform estimation without the preliminary estimate of the eigenvalues, which we now present. Although the method of estimation discussed above is straightforward in principle, in practice it is not, unless the compartmental matrix is small, that is, $n \leq 3$, since it usually is difficult to distinguish more eigenvalues. In what follows, we discuss a practical method of estimation that avoids the computation of eigenvalues and on its own could be used as a method to obtain a possibly good starting guess for a more standard procedure such as least squares.

Suppose we let our preliminary estimate now be

$$\hat{g}_2(t) = \sum_{i=1}^{m-1} y_i \delta(t - t_i) \Delta(t_i), \quad \Delta(t_i) = t_{i+1} - t_i. \tag{22.10}$$

The integral of this is a Riemann sum giving the area under the curve of height y_i and weight $\Delta(t_i)$ for each data point i. Then, upon forming the Laplace transform of (22.10), we obtain

$$\hat{G}_2(s) = \sum_{i=1}^{m-1} e^{-st_i} y_i \Delta(t_i). \tag{22.11}$$

Note that (22.11) is, except for the noise, a Riemann sum consisting of weighted averages of the data. The Laplace transform $\Phi(s)$ can then be equated, for a discrete set of s's, to (22.11). One then can proceed to estimate the parameters in the least squares sense; that is, we solve for the unknown entries of A as follows. For a chosen sequence $\{s_q\}_{q=1}^{p} > 0$, we solve

$$\min_{A=[a_{kj}]} \sum_{q=1}^{p} |\hat{G}_2(s_q) - \Phi(s_q)|^2 \tag{22.12}$$

subject to A being a compartmental matrix of the appropriate model.

The choice of the sequence $\{s_q\}$ is not critical in the sense that both $\hat{G}_2(s)$ and $\Phi(s)$ are analytic for $\mathrm{Re}(s) > 0$ and, hence, should be approximable by discrete data in principle. However, to avoid ill-conditioning, the spacing of the s_q's should be large enough so that the sum (22.12) differs considerably for different values of s_q. We start with $s_1 = 0$ and choose s_p so large that $\hat{G}_2(s_p) < \epsilon$ for some prescribed error. The other s_q's are then chosen to be equally spaced, with the number p at least as great as the number of data points so as to avoid having an under determined system of equations.

Other possibilities exist for picking the sequence which may include negative values of s. For instance, the s's could be chosen in terms of the eigenvalues of the approximation to the compartmental matrix per iteration. Alternately, the diagonal elements of the compartmental matrix, which are often a fairly good approximation to the spectrum of A, could be used. Nonetheless, notice that

$$\hat{G}_2(s) = \sum_{i=1}^{m-1} e^{-st_i} \phi(t_i)\Delta(t_i) + \sum_{i=1}^{m-1} e^{-st_i} \epsilon_i \Delta(t_i)$$
$$= \Phi_\Delta(s) + \epsilon(s),$$

where, as we previously observed, $\Phi_\Delta(s)$ is the Riemann sum of the integral defining $\Phi(s)$ and the ϵ_i are independent noise variables with zero mean. The other term, $\epsilon(s)$, is a random variable with mean 0 and variance

$$\sigma^2(s) = \sum_{i=1}^{m-1} e^{-2st_i} \Delta(t_i)^2 \sigma^2.$$

It is well known that Newton's or similar methods are sensitive to the starting guess of the minimizer (see Dennis and Schnabel (1983)), the result obtained by solving (22.12) may be refined by using it as a starting point for an ordinary least squares estimate and thereby possibly obtaining a better starting solution for a Newton numerical update. See the work of Contreras and Casella (1996), who give supportive evidence for the one-dimensional compartmental system and analyze this with a real data set.

None of the methods we have outlined work very well in practice. The reason is that the problem of estimating the parameters, i.e., flow rates, is an *ill-posed problem* (see Bellman and Aström (1970) or Walter (1987)) no matter how it is done. Even with perfect data, similar observations can come from very dissimilar models. However, more data, in the form of observations from more than one individual or experimental unit, may make the problem more tractable.

22.2 Statistical Estimation

Thus far, we have dealt with estimation only for a given data set on one subject or experimental unit. However, often there is not enough data on any one given subject to obtain a reasonable estimate of the flow rates; rather, there is a small set of observations on many individuals, such as is the case in experiments involving human subjects. Moreover, even if there are enough data so that conceivably it would be possible to obtain a good estimate on any one subject, the methods of estimation discussed previously do not address variability of the estimates within the subject or across subjects; that is, as is often the case in drug kinetics, there is as much interest on estimating the flow rates for the average individual in the population of interest as there is for any one individual. (See Soong (1971), Tsokos and Tsokos (1976), and Campello and Cobelli (1978), for some earlier work in the area). Thus, it is important to have estimation methods which make efficient use of all the available data and can account for variability within a particular individual and across individuals. The method which accounts for this is based on *hierarchical nonlinear models*, which we discussed next.

In what follows, we will use notation and language which is consistent with that found in the mixed-models literature (see Davidian and Giltinan (1995), for instance).

Let y_{ij} denote the jth measurement, $j = 1, \ldots, \hat{n}_i$, for the ith individual, $i = 1, \ldots, m$, taken at times t_{ij} so that we have a total of $N = \sum_{i=1}^{m} \hat{n}_i$ observations. The form of the model is assumed to be the same for all individuals although the parameters themselves may vary, so that for the

i^{th} individual, the jth measurement follows the model

$$y_{ij} = \phi(t_{ij}, A_i) + \epsilon_{ij}, \quad j = 1, 2, \ldots, \hat{n}_i, \qquad (22.13)$$

where the ϵ_{ij} is a random error term with mean 0 for the ith individual. We denote the vector form of (22.13) by using

$$\phi_i(A_i) := \begin{pmatrix} \phi(t_{i1}, A_i) \\ \vdots \\ \phi(t_{i\hat{n}_i}, A_i) \end{pmatrix}$$

to get

$$\boldsymbol{y}_i = \boldsymbol{\phi}_i(A_i) + \boldsymbol{e}_i, \qquad (22.14)$$

where the $\boldsymbol{e}_i$'s are independent random (noise) vectors.

Now, we proceed to specify a distribution for the errors themselves; that is, we will make the common (in linear regression) distributional assumption that the within-individual variation is normally distributed centered at 0 given the A_i. The variance/covariance matrix will be R_i. For example, in the uncorrelated case, $R_i = \sigma_i^2 I_{\hat{n}_i}$, where $I_{\hat{n}_i}$ is the $(\hat{n}_i \times \hat{n}_i)$ identity matrix. Often, however, in pharmacokinetics or the modeling of a drug through a system, it is assumed that R_i depends on the parameters of the mean function ϕ so that it is not a diagonal matrix.

To account for inter-individual or between-individual variability among the parameters A_i, we adopt the model which employs no use of covariate structure but, rather, can be interpreted as saying that the inter-individual variability is due entirely to unexplained phenomena

$$A_i = A + E_i, \qquad (22.15)$$

where A is the true but unknown average compartmental matrix of fixed parameters and E_i is a matrix of random effects assumed to arise from a population with mean zero and covariance matrix D. Once more, a complete characterization of the inter-individual variability requires an assumption about the distribution of the random effects E_i. By analogy with the linear mixture model (Searle et al. (1992)), we suppose $E_i \sim N(0, D)$; that is, that the individual specific regression parameters are normally distributed with mean A and variance/covariance matrix D.

Hence, our hierarchical nonlinear model given in the previous discussion is summarized as follows:

$$\boldsymbol{y}_i = \boldsymbol{\phi}(A_i) + \boldsymbol{e}_i,$$

where

$$e_i|A_i \sim N\{0, R_i(A_i)\}, \ A_i = A + E_i, \ \text{and} \ E_i \sim N(0, D). \qquad (22.16)$$

This model should enable us to estimate the parameters in our compartmental model. However, because of the nonlinearity of ϕ, it should first be linearized and then standard linear statistical estimation procedures should be used. However, this is best done on an individual basis for the particular experiment of interest. The reader is referred to Lindstrom and Bates (1990), Beal and Sheiner (1992), Davidian and Giltinan (1995), Littell et al. (1996), or Venables and Ripley (1997).

23

Complexity and Stability

A recurring problem in ecology is the relation between the complexity of an ecosystem and its stability. Most ecologists assumed the two concepts went together, i.e., greater complexity was associated with greater stability. However, May (1973) challenged this assumption and, indeed, showed that for Lotka–Volterra models of ecosystems, the opposite is sometimes true. However, he did not consider compartmental models and used only the number of nonzero flows as an indicator of complexity. For compartmental models, another approach, that we shall use, is possible.

23.1 Complexity

Most of the extensive literature on the relation between the complexity of a model ecosystem and various other properties uses the same definition of complexity. This is true whether the model is in the form of a general Lotka–Volterra differential equation, a donor-controlled compartmental model, or a food web described qualitatively. The index of complexity used is the average number of flows per species represented in the model. Although this is a plausible approach, it tends to count marginally important flows as much as principal flows and thus changes considerably if the former are ignored. An index in which weaker flows are given less weight would ameliorate this difficulty. It may also help clarify the relation between stability and complexity of ecosystems.

The situation is analogous to the measurement of diversity of a community , with which it is sometimes confused. Large systems may be highly diverse without being very complex, i.e., without having a very complex system of interactions. In the context of compartmental models, the diversity is a property of the number and equilibrium levels of the compartments, whereas the complexity depends on the number of flows and magnitudes of flow rates.

There is an extensive literature on diversity and various indices which have been introduced to measure it. Pielou (1969) lists three desirable properties of a diversity index, two of which are as follows:

(1) For a given number of species, the index shall have its greater value when the community is completely even.

(2) Given two completely even communities, the index should have a larger value for the one with more species.

Some of the indices used which satisfy these properties are the species count, the Simpson index, and the Shannon index. The last is based on the measure of *entropy* of a probability distribution and is related to the amount of information in a coded message. A general approach to diversity indices which includes these as well as others has been formulated by Taillie (1977).

An index of complexity should have similar properties expressed in terms of the structure, viz.:

(1) For a given structure, the index should have its greatest value when the flow rates are even (all the same).

(2) Given two structures with the same number of compartments and even flow rates, the index should have a larger value for the one with more nonzero flow rates.

We shall introduce a number of indices of complexity for compartmental models which satisfy these as well as other plausible properties to be mentioned below. One will correspond to the Shannon index; the others will be analogous to other measures of entropy.

The analogy to entropy in the case of complexity indices is even more direct than for diversity indices. Suppose that at time t, there are N molecules (or calories, or dollars, etc.) in compartment i. Each such molecule must move to an adjacent compartment or stay in the same one at time $t + 1$. The move must, of course, be via an arc emanating from compartment i. Let Y_1 be the arc chosen by molecule 1, Y_2 the one chosen by molecule 2, ..., Y_N the one chosen by molecule N. Each Y_k is a random variable with n possible values corresponding to the n compartments. Hence, the sequence

$$(Y_1, Y_2, \ldots, Y_N)$$

can be considered a coded message of length N with n symbols. The various measures of entropy used in information theory assign a number to this message which will satisfy conditions (1) and (2) for a complexity index.

Another intuitive meaning of entropy, in addition to the amount of information in a code, is as an amount of uncertainty. This interpretation also corresponds to complexity. A more complex system has a greater degree of uncertainty associated with the choice of paths for each molecule. For the least-complex model, there is no choice at all and hence zero uncertainty. The uncertainty is greatest when there are equally weighted paths (arcs) to all compartments.

Many other properties of entropy measures correspond to intuitively desirable properties of complexity indices (Taillie, 1977, p. 32). They will be shared by our indices as well.

In this chapter, we shall first explore the properties of an index analogous to the Shannon index and show it to have a simple relation to the eigenvalues of the system.

We shall assume we have a closed, linear, donor-controlled compartmental model given by a system of differential equations

$$\frac{d\mathbf{x}}{dt} = A\mathbf{x}.$$

We shall also use a discrete version written as

$$\mathbf{x}_{n+1} = (I + hA)\mathbf{x}_n,$$

which, if h is sufficiently small $[h < (\max |a_{ii}|)^{-1}]$ and $\mathbf{x}_n$ is a probability vector, becomes the defining relation of a Markov chain with transition matrix

$$P = I + hA$$

(see Chapter 12).

Although these properties and many of the definitions which follow are applicable to all closed models, most of our results will require that in addition the associated digraph be strongly connected.

In defining a complexity index, we shall, as was indicated above, use diversity indices or entropy as a guide. Each diversity index assigns a measure to the probability associated with each species, called its rarity, and the index itself is a weighted mean of these rarities. The corresponding concept for a transition probability in a compartmental model is the meagerness of a flow. The complexity of a single compartment will be the average meagerness of flows leaving that compartment.

The measure of meagerness should assign a lower complexity to compartments with meager flow than to those with stronger ones. We shall choose a family of measures which does just that. The complexity of a compartmental model will then be the sum of the complexities of the individual compartments.

DEFINITION 23.1   Let p be a nonzero transition probability, β a real number. The meagerness of p will be a monotone function of the form

$$\omega_\beta(p) = \frac{1 - p^\beta}{\beta}, \quad \beta \neq 0, \quad 0 < p \leq 1,$$

$$\omega_0(p) = -\log p.$$

DEFINITION 23.2   Let the jth compartment have as its transition probabilities p_{ij}, $i = 1, 2, \ldots, n$ (i.e., the elements in the jth column of P). Its
complexity $\Gamma_\beta(j)$ is given by

$$\Gamma_\beta(j) = \lim_{h \to 0+} \sideset{}{'}\sum_i \frac{p_{ij}\omega_\beta(p_{ij})}{h\omega_\beta(h)},$$

where $\sum'$ means that the sum is taken over all i such that $p_{ij} \neq 0$. The
β-complexity index of the compartmental model is given by

$$\Gamma_\beta = \sum_j \Gamma_\beta(j).$$

REMARK 23.1   The index Γ_0 is analogous to the familiar Shannon index
of diversity or of information theory. The others correspond, although not
exactly, to the information of other orders. For $\beta = -1$, the index counts
the number of nonzero flows in the model and thus reduces to the usual
index of complexity.

23.2   The Shannon Index

In this section, we derive a formula for the Shannon index Γ_0 and express
it in terms of the eigenvalues of the model.

THEOREM 23.2
Let A be a closed compartmental matrix. Then, the Shannon index Γ_0 of
complexity is given by

$$\Gamma_0 = -\operatorname{Tr} A = \sum_i \sum_{j \neq i} a_{ij},$$

where $\operatorname{Tr} A$ is the trace of A.

This is the Sum of all the rate Constants in the system

PROOF   The complexity of compartment j is

$$\Gamma_0(j) = \lim_{h \to 0} \sideset{}{'}\sum_i \frac{p_{ij} \log p_{ij}}{h \log h}$$

$$= \lim_{h \to 0} \sideset{}{'}\sum_i \frac{(\delta_{ij} + ha_{ij})\log(\delta_{ij} + ha_{ij})}{h \log h}, \qquad (23.1)$$

since $p_{ij} = \delta_{ij} + ha_{ij}$, where δ_{ij} is the Kronecker delta. For $i = j$, we have

$$\log(\delta_{jj} + ha_{jj}) = \log(1 + ha_{jj}) = ha_{jj} + O(h^2), \qquad (23.2)$$

whereas for $i \neq j$,

$$\log(\delta_{ij} + ha_{ij}) = \log(ha_{ij}) = \log h + \log a_{ij}. \qquad (23.3)$$

By substituting (23.2) and (23.3) into (23.1), we obtain

$$
\begin{aligned}
\Gamma_0(j) &= \lim_{h \to 0} \left[\frac{(1 + ha_{ij})[ha_{jj} + O(h^2)]}{h \log h} - \sideset{}{'}\sum_{i \neq j} \frac{ha_{ij}(\log h + \log a_{ij})}{h \log h} \right] \\
&= \lim_{h \to 0} \frac{a_{ij} + O(h)}{\log h} + \sideset{}{'}\sum_{i \neq j} a_{ij} + \lim_{h \to 0} \sideset{}{'}\sum_{i \neq j} a_{ij} \frac{\log a_{ij}}{\log h} \\
&= \sideset{}{'}\sum_{i \neq j} a_{ij}. \qquad (23.4)
\end{aligned}
$$

Since the columns of A sum to 0, $\sum_{i \neq j} a_{ij} = -a_{jj}$, and, hence, we have $\Gamma_0(j) = -a_{jj}$. We then sum over all j to obtain

$$\Gamma_0 = -\sum_{j=1}^{n} a_{jj} = -\operatorname{Tr} A.$$

∎

REMARK 23.3 If compartment j is a *trap*, i.e., has no flows leaving it, the jth column of A is all zeros. The corresponding column of P is the same as the jth column of an $n \times n$ identity matrix. Then,

$$\Gamma_0(j) = \lim_{h \to 0+} \frac{1 \log 1}{h \log h} = 0,$$

which agrees with $\Gamma_0(j) = -a_{jj} = 0$. It also agrees with the intuitive notion that the complexity in such a case should be zero.

REMARK 23.4 Since $\operatorname{Tr} A$ is invariant under similarity transformations and since the trace of the Jordan normal form of A is the sum of the eigenvalues, it follows that $\Gamma_0 = -\Sigma \lambda_i$, where λ_i are the nonzero eigenvalues of A.

23.3 Other Indices

In this section, we derive formulas for the other indices Γ_β similar to those of the Shannon index Γ_0.

THEOREM 23.5
Let A be a closed compartmental matrix. Then, the complexity index Γ_β is given by

$$\Gamma_\beta = -(\beta + 1)\operatorname{Tr} A \quad \text{for } \beta > 0,$$

by

$$\Gamma_\beta = \sum_j \sum_{i \neq j}{}' a_{ij}^{1+\beta} \quad \text{for } \beta < 0, \ \beta \neq -1,$$

and by $\Gamma_{-1} = N$ (total number of flows in the system).

PROOF As in Theorem 23.2, we have, for $\beta > 0$,

$$\Gamma_\beta(j) = \lim_{h \to 0} \sum_i{}' \frac{\delta_{ij} + ha_{ij} - (\delta_{ij} + ha_{ij})^{\beta+1}}{\beta h(1 - h^\beta)/\beta}$$

$$= \lim_{h \to 0} \frac{1 - \sum_i{}'(\delta_{ij} + ha_{ij})^{\beta+1}}{h(1 - h^\beta)}$$

$$= \lim_{h \to 0} \frac{1 - (1 + ha_{jj})^{\beta+1} - h^{\beta+1}\sum_{i \neq j}{}' a_{ij}^{\beta+1}}{h(1 - h^\beta)}$$

$$= \lim_{h \to 0} \frac{-(\beta + 1)(1 + ha_{jj})^\beta a_{jj} - (\beta + 1)h^\beta \sum_{i \neq j}{}' a_{ij}^{\beta+1}}{1 - (\beta + 1)h^\beta}$$

$$= \frac{-(\beta + 1)a_{jj} - 0}{1 - 0} = -(\beta + 1)a_{jj}. \tag{23.5}$$

Hence, $\Gamma_\beta = -(\beta + 1)\sum_j a_{jj} = -(\beta + 1)\operatorname{Tr} A$.

For $\beta < 0$ and $\beta \neq -1$, we have

$$\Gamma_\beta(j) = \lim_{h \to 0} \left[\frac{1 - (1 + ha_{jj})^{\beta+1}}{h(1 - h^\beta)} - \frac{h^{\beta+1}}{h(1 - h^\beta)} \sum_{i \neq j}{}' a_{ij}^{\beta+1} \right]$$

$$= \lim_{h \to 0} \left(\frac{-(\beta + 1)(1 + ha_{jj})^\beta a_{jj}}{1 - (\beta_1)h^\beta} - \frac{1}{h^{-\beta} - 1} \sum_{i \neq j}{}' a_{ij}^{\beta+1} \right)$$

$$= 0 + \sum_{i \neq j}{}' a_{ij}^{\beta+1}. \tag{23.6}$$

For $\beta = -1$, we find that

$$\Gamma_{-1}(j) = 0 + \lim_{h \to 0} -\frac{1}{h(1 - h^{-1})} \sum_{i \neq j}{}' a_{ij}^0$$

$$= \sum_{i \neq j}{}' a_{ij}^0 = n_j. \tag{23.7}$$

This last expression is the number of arcs leaving the jth compartment.

∎

REMARK 23.6 The complexity indices Γ_β for $\beta > 0$ are, except for a constant factor, the same as the Shannon index Γ_0 and, hence, not very interesting. The indices for $\beta < -1$ weight the weak flows more than the strong ones and, hence, are not very plausible measures of complexity. Thus, only the Γ_β for $-1 \leq \beta \leq 0$ will be considered.

REMARK 23.7 The index Γ_{-1}, which counts the number of arcs in the digraph, has the opposite property to the Shannon index Γ_0 with respect to a reduction in the number of arcs. Γ_{-1} will decrease as the number of arcs is reduced, while it may be shown (Walter (1984)) that Γ_0 will increase. This seems to indicate that a choice of Γ_β with β intermediate between 0 and -1 is appropriate.

PROPOSITION 23.8
Let $-1 < \beta < 0$ and $\alpha = -\beta$. Then,

$$\Gamma_\beta \leq (\Gamma_0)^{1-\alpha}(\Gamma_{-1})^\alpha.$$

PROOF By Theorem 23.5, $\Gamma_\beta(j) = \sum'(a_{ij})^{1-\alpha}$, from which, by Hölder's inequality, we obtain

$$\Gamma_\beta(j) \leq \left[\sum_i{}' (a_{ij})^{(1-\alpha)p}\right]^{1/p} \left[\sum_i{}' 1^q\right]^{1/q}. \tag{23.8}$$

By taking $p = 1/(1 - \alpha)$ and $q = 1/\alpha$, we find

$$\Gamma_\beta(j) \leq \left(\sum_i{}' a_{ij}\right)^{1-\alpha} n_j^\alpha, \tag{23.9}$$

where n_j is the number of arcs leaving the jth vertex. Then,

$$\Gamma_\beta = \sum_j \Gamma_\beta(j) \leq \sum_j \left(\sum_i{}' a_{ij}\right)^{1-\alpha} n_j^\alpha$$

$$\leq \left(\sum_j \sum{}' a_{ij}\right)^{1-\alpha} \left(\sum n_j\right)^\alpha = (\Gamma_0)^{1-\alpha} N^\alpha \qquad (23.10)$$

by another application of Hölder's inequality. ∎

REMARK 23.9 One of the properties we would expect a complexity index to have is a higher value when a new arc is added to the digraph but the model otherwise remains unchanged. This is clearly the case for all of the indices considered, since it merely involves adding a positive term to the sum.

REMARK 23.10 Another plausible property is for the index to be larger when the flow rates are more nearly even. This occurs when a transfer from a higher to a lower flow rate occurs.

DEFINITION 23.3   Let A be a closed compartmental matrix, and let a_1 and a_2 be two flow rates from vertex j (i.e., two off-diagonal elements in the same column j of A) such that $0 < 2h \leq a_2 - a_1$; let $a_1' = a_1 + h$, $a_2' = a_2 - h$, and let A' be the matrix with the new values in the corresponding position but otherwise the same as A. Then, A' will again be a closed compartmental matrix for any such h. It will be called a redistribution *of A.*

REMARK 23.11 The purpose of such a redistribution is to even out the flow rates. By means of a sequence of such redistributions (with different values of h), the flow rates from each vertex can be made completely even. However, the diagonal elements remain unchanged, and unless these are equal, flow rates in one column will differ from those in another.

PROPOSITION 23.12
Let B be obtained from A by a sequence of redistributions. Then, $\Gamma_\beta(B) \geq \Gamma_\beta(A)$ for $-1 \leq \beta \leq 0$.

PROOF The inequality is trivially true for $\beta = 0$ and $\beta = -1$, since in those cases, the redistribution leaves the index unchanged. For $-1 < \beta < 0$, the index is

$$\Gamma_\beta = \sum_j \sum_i {}' a_{ij}^{\beta-1}.$$

Since $x^{\beta+1}$ is a concave function for $0 < 1 + \beta < 1$, it follows that

$$a_1^{1+\beta} + a_2^{1+\beta} \leq (a_1 + h)^{1+\beta} + (a_2 - h)^{1+\beta}.$$

Hence, the redistribution results in $\Gamma_\beta(A') \geq \Gamma_\beta(A)$. ∎

COROLLARY 23.13

Let A be a closed compartmental matrix; let B be a matrix of a model with the same structure and the same main diagonal but with all flow rates from each vertex the same. Then,

$$\Gamma_\beta(B) \geq \Gamma_\beta(A)$$

for $-1 < \beta < 0$, with equality holding only if $B = A$.

REMARK 23.14 The most complex model with a given structure occurs when all the nonzero flow rates are the same. If the structures changes as well, the complexity increases with an increase in the number of arcs. Hence, the maximum complexity for each index occurs when each pair of vertices are joined by arcs and all flow rates are the same.

23.4   Relation to Stability

The relation between the Shannon index and the eigenvalues of the matrix has already been noted. It is given by

$$\Gamma_0 = -\sum \lambda_i,$$

where λ_i are the nonzero eigenvalues of A. Moreover, each of the other indices is majorized by a geometric mean of Γ_0 and Γ_{-1}; therefore,

$$\Gamma_{-\alpha} \leq \left(\sum(-\lambda_i)\right)^{1-\alpha} N^\alpha, \quad 0 < \alpha < 1, \tag{23.11}$$

where N is the total number of arcs. Greater relative stability is usually associated with eigenvalues whose real part is highly negative. Since

$$-\sum \lambda_i = \sum(-\operatorname{Re}\lambda_i),$$

Γ_0 should be larger for systems with greater stability, and conversely. The converse also holds for $\Gamma_{-\alpha}$, which, if it is large, ensures that the real parts of all eigenvalues have a large negative value.

The relation between these complexity indices and the resilience index introduced in Walter (1983) can also be examined. The latter was shown to be equivalent to

$$r = -\sum_{i=2}^{n} \lambda_i^{-1}$$

for strongly connected compartmental models(SCCM). (Such models have at most one zero eigenvalue.) The same conclusions reached in that work for the index c (proportional to our Γ_{-1}) hold here. In fact, Γ_0 and r are approximate reciprocals if the eigenvalues are close to each other. In the work of Walter (1980), it was shown that $\Gamma_{-1} \leq \frac{1}{m} \sum -\lambda_i$ where m is the minimum flow rate a_{ij}, $i \neq j$. Hence, from (23.11), we have

$$\Gamma_{-\alpha} \leq \left(\sum(-\lambda_i)\right)^{1-\alpha} m^{-\alpha} \left(\sum(-\lambda_i)\right)^{\alpha}$$
$$= m^{-\alpha} \sum(-\lambda_i).$$

Thus, we may conclude that a large $\Gamma_{-\alpha}$ forces r to be small in the presence of similar eigenvalues. Since a small r corresponds to greater resilience, it will be associated with greater complexity.

However, models with the same complexity indicex Γ_β may have different ranges of eigenvalues. The two examples in Figure 23.1 have the same Γ_β, but in one case, the resilience index $r = \frac{6}{13}$, whereas in another it is $r = \frac{6}{5}$.

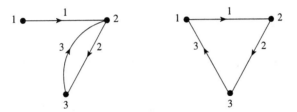

Figure 23.1: Compartmental models with the same complexity and flow rates, but with different structure.

The question remains as to whether the eigenvalues have a large spread. This was already taken up in Chapter 20, in which the spread of the eigenvalues was given in terms of the structure. For catenary and mammillary systems, the spread depends on the evenness of the flow rates. This spread, the sum of squares deviation about the mean, is partitioned into the sum of squares due to connections of the cycles and between the cycles. The latter is zero if each cycle has the same eigenvalue. In such a case, the complexity and resilience indices behave the same way.

A

Mathematical Prerequisites

We collect here a number of mathematical techniques that are used in the body of the book. Elements from matrix theory and from differential equations are included. These are much abbreviated versions of results found in any good book on matrix theory or systems of differential equations (e.g. Rabenstein (1982), Borrelli and Coleman (1987), Marcus and Minc (1964)).

A.1 Matrix Operations

A matrix A is a rectangular array of numbers of the form

$$A = \begin{bmatrix} a_{11} & a_{12} & \cdots & a_{1n} \\ a_{21} & a_{22} & \cdots & a_{2n} \\ \cdots & \cdots & & \cdots \\ a_{m1} & a_{m2} & \cdots & a_{mn} \end{bmatrix} \tag{A.1}$$

with m rows and n columns. Such an array is called an $m \times n$ matrix. This array uses the standard convention that a_{ij} is the number in the ith row and j^{th} column. These individual numbers a_{ij} are called the *elements* of the matrix A, and are, as we have done here, usually denoted by lowercase letters. The entire matrix is usually represented by a single capital letter, but sometimes is denoted by $A = [a_{ij}]$ rather than the expression in (A.1).

A matrix may be multiplied by a number c (called a scalar) and two matrices of the same size may be added. For example, if we let

$$A = \begin{bmatrix} 1 & 2 & 3 \\ 6 & 5 & 4 \end{bmatrix}, \quad B = \begin{bmatrix} 0 & 1 & 1 \\ 1 & -1 & 0 \end{bmatrix},$$

then

$$A + B = \begin{bmatrix} 1 & 3 & 4 \\ 7 & 4 & 4 \end{bmatrix}, \quad 2A = \begin{bmatrix} 2 & 4 & 6 \\ 12 & 10 & 8 \end{bmatrix}.$$

These two operations satisfy all the properties inherited from addition and multiplication of numbers. Matrices that are not the same size cannot be added. However, for the other important matrix operation, multiplication, they do not need to be the same size.

Matrix multiplication is based on vector dot products. In fact, vectors are just special matrices. A *row vector* is a $1 \times n$ matrix, whereas a *column vector* is an $n \times 1$ matrix. For example, $B = \begin{bmatrix} b_1 \\ b_2 \end{bmatrix}$ is a 2×1 column vector and $C = [c_1 \ c_2 \ c_3]$ is a 1×3 row vector. For $n = 3$, either can be used to represent geometric vectors in three-dimensional space. The dot product of such vectors is the same as the matrix product of a $1 \times n$ row vector with a $n \times 1$ column vector, e.g.,

$$[2 \ \ 1 \ \ 3] \begin{bmatrix} 1 \\ 0 \\ -1 \end{bmatrix} = 2 \cdot 1 + 1 \cdot 0 + 3 \cdot (-1) = -1.$$

An $m \times n$ matrix may be multiplied by an $n \times 1$ column matrix by multiplying each of the rows by this column using the same rule, i.e.,

$$\begin{bmatrix} 1 & 2 & 3 \\ 6 & 5 & 4 \end{bmatrix} \begin{bmatrix} 1 \\ 0 \\ -1 \end{bmatrix} = \begin{bmatrix} 1 \cdot 1 + 2 \cdot 0 + 3 \cdot (-1) \\ 6 \cdot 1 + 5 \cdot 0 + 4 \cdot (-1) \end{bmatrix} = \begin{bmatrix} -2 \\ 2 \end{bmatrix}.$$

This rule may be extended to the product of $m \times n$ matrices by $n \times p$ matrices. Thus, we get

$$\begin{bmatrix} 1 & 2 & 3 \\ 6 & 5 & 4 \end{bmatrix} \begin{bmatrix} 1 & 0 & 1 \\ 0 & 1 & -1 \\ -1 & 1 & 0 \end{bmatrix} = \begin{bmatrix} -2 & 5 & -1 \\ 2 & 9 & 1 \end{bmatrix}.$$

This multiplication rule may be formalized for any $m \times n$ matrix A and $n \times p$ matrix B to get the product $C = AB$. If $A = [a_{ij}]$ and $B = [b_{ij}]$, then the product $C = [c_{ij}]$ is given by

$$c_{ij} = \sum_{k=1}^{n} a_{ik} b_{kj}, \quad i = 1, \ldots, m, j = 1, \ldots, p. \tag{A.2}$$

The properties of matrix multiplication are *not* the same as those of multiplication of numbers. The product does satisfy the associative and distributive laws:

$$(AB)C = A(BC),$$
$$A(B + C) = AB + AC, \tag{A.3}$$
$$(A + B)C = AC + BC,$$

provided the matrices are the right size. But

$$AB \neq BA,$$

except in special circumstances.

The numbers 0 and 1 correspond to the $m \times n$ matrix O and $n \times n$ matrix I:

$$O = \begin{bmatrix} 0 & 0 & \cdots & 0 \\ 0 & 0 & \cdots & 0 \\ \vdots & \vdots & & \vdots \\ 0 & 0 & \cdots & 0 \end{bmatrix}, \quad I = \begin{bmatrix} 1 & 0 & \cdots & 0 \\ 0 & 1 & \cdots & 0 \\ \vdots & \vdots & & \vdots \\ 0 & 0 & \cdots & 1 \end{bmatrix},$$

which are respectively the identities of addition and multiplication.

There are also some operations on single matrices. These incude the *transpose*, the *trace*, and the *determinant* of a matrix. The transpose, A^T, involves interchanging the rows and columns of the matrix A. For example,

$$\begin{bmatrix} 1 & 2 & 3 \\ 6 & 5 & 4 \end{bmatrix}^T = \begin{bmatrix} 1 & 6 \\ 2 & 5 \\ 3 & 4 \end{bmatrix}.$$

The trace of a square $(n \times n)$ matrix A, denoted $\mathrm{Tr}(A)$, is the sum of all its diagonal elements. If $A = [a_{ij}]$, then $\mathrm{Tr}(A) = a_{11} + a_{22} + \cdots + a_{nn}$:

$$\mathrm{Tr}\left(\begin{bmatrix} 1 & 2 \\ 4 & 3 \end{bmatrix}\right) = 1 + 3 = 4.$$

The determinant, $\det(A)$, of a triangular matrix A, i.e., a square matrix with all zeros either above or below the main diagonal, is the product of its diagonal elements. For example,

$$\det\left(\begin{bmatrix} 1 & 2 \\ 0 & 3 \end{bmatrix}\right) = 1 \cdot 3 = 3, \quad \det\left(\begin{bmatrix} 1 & 0 & 0 \\ 2 & 3 & 0 \\ 3 & 4 & 5 \end{bmatrix}\right) = 1 \cdot 3 \cdot 5 = 15.$$

The definition of determinant may be extended to nontriangular matrices, but in this case, it is *not* the same as this product. This involves using elementary row operations to reduce a matrix to triangular form.

The first of these elementary row operations involves interchanging two rows of the matrix. The effect on the determinant is to change the sign:

$$\det\left(\begin{bmatrix} 0 & 3 \\ 1 & 2 \end{bmatrix}\right) = -3.$$

The second kind of elementary row operations involves adding a multiple of one row to another. It doesn't change the determinant. For example, if we add two times the third row of the matrix

$$\begin{bmatrix} 1 & 0 & 0 \\ 2 & 3 & 0 \\ 3 & 4 & 5 \end{bmatrix}$$

to the first row, we get

$$\begin{bmatrix} 7 & 8 & 10 \\ 2 & 3 & 0 \\ 3 & 4 & 5 \end{bmatrix},$$

whose determinant is still 15.

Now, we reverse the procedure to find the determinant of any square matrix. A succession of the two types of operations will reduce it to triangular form. For example, the following equation reduces the matrix:

$$\begin{bmatrix} 1 & 2 & 3 \\ 6 & 5 & 4 \\ 7 & 8 & 9 \end{bmatrix} \rightarrow \begin{bmatrix} 1 & 2 & 3 \\ 0 & -7 & -14 \\ 7 & 8 & 9 \end{bmatrix} \rightarrow \begin{bmatrix} 1 & 2 & 3 \\ 0 & -7 & -14 \\ 0 & -6 & -12 \end{bmatrix}$$

$$\rightarrow \begin{bmatrix} 1 & 2 & 3 \\ 0 & -7 & -14 \\ 0 & 0 & 0 \end{bmatrix}.$$

The first operation adds (-6) times the first row to the second, the second operation adds (-7) times the first row to the third, and the last operation adds $(-6/7)$ times the second row to the third. None of these has changed the determinant, so that the determinant of the matrix is 0.

Here we have used only the second type of elementary row operations. We may have to use the first type if, at some stage, a zero appears on the main diagonal. For example, the matrix

$$\begin{bmatrix} 1 & 2 & 3 \\ 2 & 4 & 5 \\ 3 & 5 & 6 \end{bmatrix} \longrightarrow \begin{bmatrix} 1 & 2 & 3 \\ 0 & 0 & -1 \\ 3 & 5 & 6 \end{bmatrix} \longrightarrow \begin{bmatrix} 1 & 2 & 3 \\ 0 & 0 & -1 \\ 0 & -1 & -3 \end{bmatrix},$$

which can be put into triangular form by interchanging the second and third row. Thus, the determinant of this matrix is $-(1)(-1)(-1) = -1$.

This same procedure is used to solve linear equations. In this case, it no longer is applied only to square matrices. The linear equation problem is to find a column vector $\boldsymbol{x}$ such that

$$A\boldsymbol{x} = \boldsymbol{b},$$

where A is an $m \times n$ matrix, $\boldsymbol{x}$ a $n \times 1$ matrix, and $\boldsymbol{b}$ an $m \times 1$ matrix. The same elementary row operations are applied to both A and $\boldsymbol{b}$. For example, a solution to

$$\begin{bmatrix} 1 & 2 & 3 \\ 2 & 4 & 5 \\ 3 & 5 & 6 \end{bmatrix} \begin{bmatrix} x_1 \\ x_2 \\ x_3 \end{bmatrix} = \begin{bmatrix} 1 \\ 0 \\ -1 \end{bmatrix}$$

is found by reducing the matrices to

$$\begin{bmatrix} 1 & 2 & 3 \\ 0 & -1 & -3 \\ 0 & 0 & -1 \end{bmatrix} \begin{bmatrix} x_1 \\ x_2 \\ x_3 \end{bmatrix} = \begin{bmatrix} 1 \\ -4 \\ -2 \end{bmatrix},$$

where $\begin{bmatrix} 1 \\ -4 \\ -2 \end{bmatrix}$ is obtained from $\begin{bmatrix} 1 \\ 0 \\ -1 \end{bmatrix}$ by the same sequence of operations. Then, $x_3 = 2$ and this may be used in the second equation to find

$$-x_2 - 3 \cdot 2 = -4, \quad \text{or } x_2 = -2,$$

and then both are used to find x_1,

$$x_1 + 2(-2) + 3(2) = 1, \quad \text{or } x_1 = -1.$$

The same technique also enables us to find an inverse of a square matrix when it exists. For a *diagonal* matrix, the inverse is easy to find: If

$$D = \begin{bmatrix} d_1 & 0 & \cdots & 0 \\ 0 & d_2 & \cdots & 0 \\ \vdots & \vdots & & \vdots \\ 0 & 0 & \cdots & d_n \end{bmatrix}$$

and $\det(D) \neq 0$, then D^{-1} is

$$D^{-1} = \begin{bmatrix} 1/d_1 & 0 & \cdots & 0 \\ 0 & 1/d_2 & \cdots & 0 \\ \vdots & \vdots & & \vdots \\ 0 & 0 & \cdots & 1/d_n \end{bmatrix}$$

Clearly, $DD^{-1} = D^{-1}D = I$, but D^{-1} does not exist if any diagonal element $= 0$. We now use the elementary row operation to solve $AB = I$, where A is a given $n \times n$ matrix. The procedure is to use these operations to reduce A to diagonal form, apply the sample operations to I, and then to find the inverse of this diagonal matrix. For example, if we form the augmented matrix by combining A and I into a single matrix, we get

$$A \mid I = \begin{bmatrix} 1 & 2 & 3 & | & 1 & 0 & 0 \\ 2 & 4 & 5 & | & 0 & 1 & 0 \\ 3 & 5 & 6 & | & 0 & 0 & 1 \end{bmatrix}$$

$$\rightarrow \begin{bmatrix} 1 & 2 & 3 & | & 1 & 0 & 0 \\ 0 & 0 & -1 & | & -2 & 1 & 0 \\ 0 & -1 & -3 & | & -3 & 0 & 1 \end{bmatrix}$$

$$\rightarrow \begin{bmatrix} 1 & 2 & 3 & | & 1 & 0 & 0 \\ 0 & -1 & -3 & | & -3 & 0 & 1 \\ 0 & 0 & -1 & | & -2 & 1 & 0 \end{bmatrix}$$

$$\rightarrow \begin{bmatrix} 1 & 2 & 0 & | & -5 & 3 & 0 \\ 0 & -1 & 0 & | & 3 & -3 & 1 \\ 0 & 0 & -1 & | & -2 & 1 & 0 \end{bmatrix}$$

$$\rightarrow \begin{bmatrix} 1 & 0 & 0 & \bigg| & 1 & -3 & 2 \\ 0 & -1 & 0 & \bigg| & 3 & -3 & 1 \\ 0 & 0 & -1 & \bigg| & -2 & 1 & 0 \end{bmatrix}.$$

Now, we need merely solve the equation

$$\begin{bmatrix} 1 & 0 & 0 \\ 0 & -1 & 0 \\ 0 & 0 & -1 \end{bmatrix} B = \begin{bmatrix} 1 & -3 & 2 \\ 3 & -3 & 1 \\ -2 & 1 & 0 \end{bmatrix}$$

by inverting the diagonal matrix on the left to get

$$B = \begin{bmatrix} 1 & -3 & 2 \\ -3 & 3 & -1 \\ 2 & -1 & 0 \end{bmatrix}.$$

This procedure fails only if the inverse does not exist.

A.2   Finding Eigenvalues and Eigenvectors

If a square matrix does not have an inverse, it is said to be *singular*. Such singular matrices have zero determinant and conversely, since the same elementary row operations are used in both inverse and determinant calculations. If the determinant is 0, the triangular matrix obtained from our elementary row operations must have a zero on the main diagonal. But, then, it is not invertible.

Any square matrix can be made into a singular matrix by subtracting a multiple of the identity matrix, $A - \lambda I$. Those values of λ that work are called *eigenvalues*. They are found by finding the zeros of the *characteristic polynomial*

$$P(\lambda) = \det(A - \lambda I).$$

Since this is a polynomial of degree n, it has at most n zeros, which may be real or complex. For example, the eigenvalues of

$$A = \begin{bmatrix} 1 & 2 \\ 4 & 3 \end{bmatrix} \tag{A.4}$$

are obtained from

$$P(\lambda) = \det \left(\begin{bmatrix} 1 & 2 \\ 4 & 3 \end{bmatrix} - \lambda \begin{bmatrix} 1 & 0 \\ 0 & 1 \end{bmatrix} \right) = \det \left(\begin{bmatrix} 1 - \lambda & 2 \\ 4 & 3 - \lambda \end{bmatrix} \right)$$

$$= \lambda^2 - 4\lambda - 5 = 0$$

Thus, $\lambda = -1$ and $\lambda = 5$ are the eigenvalues. Any singular matrix A has the property that

$$A\boldsymbol{x} = \boldsymbol{0}$$

has a nonzero solution (i.e., a column vector not all of whose elements are 0). The nonzero solutions to

$$(A - \lambda I)x = 0,$$

where λ is an eigenvalue, are called *eigenvectors*. The eigenvectors belonging to different eigenvalues are distinct (i.e., linearly independent). For example, the eigenvectors corresponding to (-1) in A is the vector x which satisfies

$$\left(\begin{bmatrix} 1 & 2 \\ 4 & 3 \end{bmatrix} - (-1) \begin{bmatrix} 1 & 0 \\ 0 & 1 \end{bmatrix} \right) x = \begin{bmatrix} 2 & 2 \\ 4 & 4 \end{bmatrix} x = \begin{bmatrix} 0 \\ 0 \end{bmatrix},$$

which is

$$x = \begin{bmatrix} 1 \\ -1 \end{bmatrix}.$$

It's clear that any multiplier of this is also an eigenvector, but it must be *nonzero*. The eigenvector belonging to $\lambda = 5$ satisfies

$$\begin{bmatrix} -4 & 2 \\ 4 & -2 \end{bmatrix} x = \begin{bmatrix} 0 \\ 0 \end{bmatrix}$$

or

$$x = \begin{bmatrix} 1 \\ 2 \end{bmatrix}.$$

If there are enough eigenvectors (n linearly independent), then A has a simple decomposition. We denote by K the matrix when columns are these eigenvectors. Then,

$$AK = K\Lambda,$$

where Λ is the diagonal matrix with the eigenvalues on the main diagonal, and

$$A = K\Lambda K^{-1}.$$

This simplifies many calculations involving A. For example, A^2 is given by

$$A^2 = K\Lambda K^{-1} K\Lambda K^{-1} = K\Lambda^2 K^{-1}$$

and, similarly,

$$A^m = K\Lambda^m K^{-1}.$$

This enables us to find a simple expression for any polynomial in A,

$$p(A) = Kp(\Lambda)K^{-1},$$

and even functions of A given by power series, e.g.,

$$e^A = Ke^\Lambda K^{-1} = K \begin{bmatrix} e^{\lambda_1} & 0 & \cdots & 0 \\ 0 & e^{\lambda_2} & \cdots & 0 \\ \vdots & \vdots & & \vdots \\ 0 & 0 & \cdots & e^{\lambda_n} \end{bmatrix} K^{-1}.$$

In the example previously considered, we have

$$\begin{bmatrix} 1 & 2 \\ 4 & 3 \end{bmatrix} = \begin{bmatrix} 1 & 1 \\ -1 & 2 \end{bmatrix} \begin{bmatrix} -1 & 0 \\ 0 & 5 \end{bmatrix} \begin{bmatrix} 2/3 & -1/3 \\ 1/3 & 1/3 \end{bmatrix}$$

and, therefore,

$$\begin{bmatrix} 1 & 2 \\ 4 & 3 \end{bmatrix}^2 = \begin{bmatrix} 1 & 1 \\ -1 & 2 \end{bmatrix} \begin{bmatrix} 1 & 0 \\ 0 & 25 \end{bmatrix} \begin{bmatrix} 2/3 & -1/3 \\ 1/3 & 1/3 \end{bmatrix};$$

also we have

$$e^{\begin{bmatrix} 1 & 2 \\ 4 & 3 \end{bmatrix}} = \begin{bmatrix} 1 & 1 \\ -1 & 2 \end{bmatrix} \begin{bmatrix} e^{-1} & 0 \\ 0 & e^5 \end{bmatrix} \begin{bmatrix} 2/3 & -1/3 \\ 1/3 & 1/3 \end{bmatrix}. \tag{A.5}$$

The power series definition could also be used directly to find the exponential function. In the case of the example, this is

$$e^{\begin{bmatrix} 1 & 2 \\ 4 & 3 \end{bmatrix}} = \sum_{k=0}^{\infty} \frac{1}{k!} \begin{bmatrix} 1 & 2 \\ 4 & 3 \end{bmatrix}^k.$$

It may be that the eigenvectors of A do not constitute a basis of $\mathbf{R}^n$, i.e., that there are fewer than n linearly independent eigenvectors. In this case, the exponential matrix function is more complex. For example, the matrix given by B,

$$B = \begin{bmatrix} 1 & 1 \\ 0 & 1 \end{bmatrix},$$

has as its powers

$$B^n = \begin{bmatrix} 1 & n \\ 0 & 1 \end{bmatrix}$$

from which it follows that

$$e^B = \sum_{n=0}^{\infty} \frac{1}{n!} \begin{bmatrix} 1 & n \\ 0 & 1 \end{bmatrix} = \begin{bmatrix} e & e \\ 0 & e \end{bmatrix}.$$

The general form which replaces the diagonal matrix Λ is the Jordan normal form for which we refer the interested reader to the references mentioned previously.

A.3 Systems of Differential Equations

A first-order linear differential equation

$$y' = ay + b$$

with initial condition $y(0) = y_0$ has a simple solution. If a is a constant, it is given by

$$y(t) = y_0 e^{at} + \int_0^t e^{a(t-s)} b(s) ds$$

$$= y_0 e^{at} + y_p(t). \tag{A.6}$$

If, in addition, b is a constant, then for $a \neq 0$,

$$y_p(t) = \frac{b}{a}(e^{at} - 1).$$

Thus, if $a < 0$, $y(t) \to -\frac{b}{a}$, i.e., is asymptotically stable, but if $a > 0$, $y(t) \to \infty$ or $-\infty$, depending on b.

If we have a system of first-order linear equations,

$$x_1' = a_{11}x_1 + a_{21}x_2 + \cdots + a_{n1}x_n + b_1,$$

$$x_2' = a_{12}x_1 + a_{22}x_2 + \cdots + a_{n2}x_n + b_2,$$

$$\vdots \tag{A.7}$$

$$x_n' = a_{1n}x_1 + a_{2n}x_2 + \cdots + a_{nn}x_n + b_n,$$

which may be written in matrix form

$$x' = Ax + b,$$

we have a solution that is similar to the single equation (A.6):

$$x(t) = e^{At} x_0 + \int_0^t e^{A(t-s)} b(s) ds. \tag{A.8}$$

Here. we have again assumed that A is a constant matrix. If b is a constant vector, then the solution is as before:

$$x(t) = e^{At} x_0 + (e^{At} - I) A^{-1} b, \tag{A.9}$$

provided A has an inverse. If A has a complete set of eigenvectors, then we may express e^{At} as

$$e^{At} = K e^{\Lambda t} K^{-1},$$

as we saw in the last section. The integral in (A.8) may be evaluated by using this form even when A is singular (but the solution (A.9) would then, of course, not be correct).

The solution to the homogeneous equation, where $b = 0$, may be expressed as

$$x_h(t) = K e^{\Lambda t} K^{-1} x_0 = K e^{\Lambda t} c$$

$$= c_1 k_1 e^{\lambda_1 t} + c_2 k_2 e^{\lambda_2 t} + \cdots + c_n k_n e^{\lambda_n t}. \tag{A.10}$$

If all of the eigenvalues are negative or have a negative real part, then $x_h(t) \to 0$ as $t \to \infty$. This is typical for compartmental models. Then, the solution (A.9) to the nonhomogeneous equation converges to $-A^{-1}b$ as $t \to \infty$.

As an example, consider, again, the matrix A of (A.4). The solution to Equation (A.7) is

$$x(t) = \frac{1}{3} \begin{bmatrix} 1 & 1 \\ -1 & 2 \end{bmatrix} \begin{bmatrix} e^{-t} & 0 \\ 0 & e^{5t} \end{bmatrix} \begin{bmatrix} 2 & -1 \\ 1 & 1 \end{bmatrix} \left\{ \begin{bmatrix} x_{01} \\ x_{02} \end{bmatrix} + \frac{1}{3} \begin{bmatrix} 2 & -1 \\ 1 & 1 \end{bmatrix} \begin{bmatrix} b_0 \\ b_1 \end{bmatrix} \right\}$$

$$- \frac{1}{3} \begin{bmatrix} 2 & -1 \\ 1 & 1 \end{bmatrix} \begin{bmatrix} b_0 \\ b_1 \end{bmatrix} \tag{A.11}$$

from (A.9) after a little algebraic manipulation. The solution to the homogeneous equation may also be expressed as

$$x_h(t) = c_1 \begin{bmatrix} 1 \\ -1 \end{bmatrix} e^{-t} + c_2 \begin{bmatrix} 1 \\ 2 \end{bmatrix} e^{5t} = \begin{bmatrix} 1 & 1 \\ -1 & 2 \end{bmatrix} \begin{bmatrix} e^{-t} & 0 \\ 0 & e^{5t} \end{bmatrix} \begin{bmatrix} c_1 \\ c_2 \end{bmatrix}. \tag{A.12}$$

In this case, one of the eigenvalues is positive, so $x_h(t)$ does not converge to 0 as $t \to \infty$.

Problem A.3.1

Find values of c_1 and c_2 in (A.12) which give the solution in (A.11) for $b_0 = b_1 = 0$.

As another example, consider the matrix

$$B = \begin{bmatrix} -5 & 2 \\ 4 & -3 \end{bmatrix}.$$

Its eigenvalues are both negative: $\lambda_1 = -7$, $\lambda_2 = -1$; but the eigenvectors are as before. The solution to the homogeneous equation

$$x' = Bx$$

is now

$$x_h(t) = c_1 e^{-7t} \begin{bmatrix} 1 \\ -1 \end{bmatrix} + c_2 e^{-t} \begin{bmatrix} 1 \\ 2 \end{bmatrix},$$

which clearly converges to 0 as $t \to \infty$.

A third example is one with the matrix

$$C = \begin{bmatrix} -4 & 2 \\ 4 & -2 \end{bmatrix}.$$

In this case, we have a singular matrix with eigenvalues 0 and -6. The solution to the homogeneous equation is

$$x_h(t) = c_1 e^{-6t} \begin{bmatrix} 1 \\ -1 \end{bmatrix} + c_2 \begin{bmatrix} 1 \\ 2 \end{bmatrix}$$

and the limit is

$$\lim_{t \to \infty} x_h(t) = c_2 \begin{bmatrix} 1 \\ 2 \end{bmatrix}.$$

The value of c_2 may be found from the initial conditions since

$$x_1(0) + x_2(0) = \begin{bmatrix} 1 & 1 \end{bmatrix} x_h(0) = c_1 \cdot 0 + c_2 \cdot 5.$$

Problem A.3.2

Find the general solution to the homogeneous equation $x' = Ax$ in the form of (A.10) for the matrices

$$A_1 = \begin{bmatrix} -2 & 1 & 0 \\ 1 & -2 & 0 \\ 0 & 0 & -2 \end{bmatrix}, \ A_2 = \begin{bmatrix} -2 & 0 & 1 \\ 1 & -2 & 0 \\ 0 & 1 & -2 \end{bmatrix}, \ A_3 = \begin{bmatrix} -1 & 0 & 1 \\ 1 & -1 & 0 \\ 0 & 1 & -1 \end{bmatrix}$$

and describe the behavior of $x_h(t)$ as $t \to \infty$.

Problem A.3.3

Find a solution to the equation

$$x' = Ax + b \quad \text{where } b = \begin{bmatrix} 1 \\ 0 \\ 0 \end{bmatrix}.$$

A.4   Matrices with Maple

Maple has a library package for linear algebra which can be used for many matrix operations. With it, we can compute eigenvalues and eigenvectors and determinants of matrices, perform row operations, and calculate inverses. Many of these procedures can be used on symbolic matrices.

To use the package, we must first read it in by typing *with(linalg);*. This also lists the names of the functions that are now available. But we'll use a ":" instead at the end to avoid listing them all.

```
> with(linalg):
```

The linalg package contains a couple of shortcut functions for creating vectors and matrices. Here's how to create a vector and a matrix:

```
> a:=vector([2,sin(x),4,5.3,beta]);
```

$$a := \begin{bmatrix} 2 & \sin(x) & 4 & 5.3 & \beta \end{bmatrix}$$

```
> A:=matrix([[1,x,y],[0,1,z],[0,0,1]]);
```

$$A := \begin{bmatrix} 1 & x & y \\ 0 & 1 & z \\ 0 & 0 & 1 \end{bmatrix}$$

Maple uses its &* operator to denote matrix multiplication. We must use the function evalm to evaluate a matrix expression.

```
> evalm(A&*A);
```

$$\begin{bmatrix} 1 & 2x & 2y + xz \\ 0 & 1 & 2z \\ 0 & 0 & 1 \end{bmatrix}$$

We could also have used the exponent operator in this case:

```
> evalm(A^2);
```

$$\begin{bmatrix} 1 & 2x & 2y + xz \\ 0 & 1 & 2z \\ 0 & 0 & 1 \end{bmatrix}$$

Maple can compute determinants:

```
> det(A);
```

$$1$$

The inverse of a matrix is given by

```
> inverse (A);
```

$$\begin{bmatrix} 1 & -x & xz - y \\ 0 & 1 & -z \\ 0 & 0 & 1 \end{bmatrix}$$

Maple even knows about the matrix exponential:

$$e^A$$

This is useful, as we have seen, in the theory of systems of linear ordinary differential equations:

```
> exponential(A);
```

$$\begin{bmatrix} e & xe & \frac{1}{2}xze + ye \\ 0 & e & ze \\ 0 & 0 & e \end{bmatrix}$$

```
> B:=matrix(2,2,[1,2,4,3]);
```

$$B := \begin{bmatrix} 1 & 2 \\ 4 & 3 \end{bmatrix}$$

Here, we have used another method of entering the matrix; the size is followed by the entries in each successive row. The exponential function that gives us the solution of a system of differential equations is easily calculated:

```
> exponential(B*t);
```

$$\begin{bmatrix} \frac{2}{3}e^{(-t)} + \frac{1}{3}e^{(5t)} & \frac{1}{3}e^{(5t)} - \frac{1}{3}e^{(-t)} \\ \frac{2}{3}e^{(5t)} - \frac{2}{3}e^{(-t)} & \frac{1}{3}e^{(-t)} + \frac{2}{3}e^{(5t)} \end{bmatrix}$$

A critical computation for such equations is also determining eigenvalues and eigenvectors:

```
> eigenvals(B);
```

$$5, -1$$

```
> eigenvects(B);
```

$$[-1, 1, \{[-1 \ \ 1]\}], [5, 1, \{[1 \ \ 2]\}]$$

We consider another numerical matrix:

```
> C:=matrix([[2,3],[4,5]]);
```

$$C := \begin{bmatrix} 2 & 3 \\ 4 & 5 \end{bmatrix}$$

The irrational eigenvalues may also be found by using the same command:

```
> eigenvals(C);
```

$$\frac{7}{2} + \frac{1}{2}\sqrt{57}, \ \frac{7}{2} - \frac{1}{2}\sqrt{57}$$

```
> charpoly(C,lambda);
```

$$\lambda^2 - 7\lambda - 2$$

Each list in this sequence contains an eigenvalue, an algebraic multiplicity, and an eigendirection. We specified the radical option here to make Maple list each eigenvector separately, rather than using Maple's RootOf notation, which initially is somewhat confusing.

```
> eigenvects(C);
```

$$\left[\text{RootOf}(\_Z^2 - 7\_Z - 2),\ 1,\ \left\{ \left[1\quad \frac{1}{3}\text{RootOf}(\_Z^2 - 7\_Z - 2) - \frac{2}{3} \right] \right\} \right]$$

Maple also has a library package for solving differential equations. It is loaded by using the command *with(DEtools);*, and may be combined with the matrix package to solve many problems in Part III.

Bibliography

[1] Anderson, D. (1983), *Compartmental Modeling and Tracer Kinetics*, Lect Notes in Biomathematics Vol. 50, Springer-Verlag, Berlin.

[2] Audoly, S., L. D'Angio, M. P. Saccomani, and C. Cobelli (1998), Global identifiability of linear compartmental models–A computer algebra algorithm, *IEEE transactions of biomedical engineering*, **45**, No. 1, 36-47.

[3] Batchelet, E., L. Brand, and A. Steiner (1979), On the kinetics of lead in the human body, *J. Math. Biol.* **8**, 15–23.

[4] Bates, D. M. and D. G. Watts (1988), *Nonlinear regression analysis and its applications*, Wiley, New York.

[5] Beal, S. L. and L. B. Sheiner (1992), *NONMEM User's Guides*, NONMEM Project Group, University of California, San Francisco.

[6] _____ (1982), Estimating population kinetics, *CRC Crit. Rev. in Biomed. Eng.* **8**, 195–222.

[7] Bellman, R. E. and R. S. Roth (1984), *The Laplace Transform*, World Scientific, Singapore.

[8] _____ and K. J. Aström (1970), On structural identifiability, *Math. Biosci.* **7**, 329–339.

[9] _____, R. E. Kalaba, and I. A. Lockett (1966), *Numerical Inversion of the Laplace Transform*, Elsevier, New York.

[10] Beltrami, E. (1993), *Mathematical Models in the Social and Biological Sciences*, Jones and Bartlett, Boston.

[11] Bender, E. (1978), *An Introduction to Mathematical Modeling*, Wiley, New York.

[12] Beverton, R. J. H. and S. J. Holt (1957), On the dynamics of exploited fish populations, *Fish. Invest. London, Series 2*, 19: 533 p.

[13] Bjorck, A. (1996), *Numerical Methods for Least Squares Problems*, SIAM, Philadelphia.

[14] Borrelli, R. L. and C. S. Coleman (1988), *Differential equations: A*

modeling approach, Prentice Hall, Englewood Cliffs, New Jersey.

[15] Campello, L. and C. Cobelli (1978), Parameter estimation of biological stochastic compartmental models–An application, *IEEE transactions of biomedical engineering* **BME-25**, No. 2, 139-146.

[16] Castillo-Chavez, C., Ed. (1989), *Mathematical and Statistical Applications in AIDS Epidemology*, Lectures Notes in Biomathematics Vol. 83, Springer-Verlag, Berlin.

[17] Chartrand, G. (1977), *Graphs as Mathematical Models*, Prindle, Weber, & Schmidt, Boston.

[18] Clark, S. H. and B. E. Brown (1977), Changes in biomass of finfish and squids from the Gulf of Maine to Cape Hatteras, 1963–1974, as determined from research vessel survey data, *U. S. Fish. Bull.* **75**, 1–21.

[19] Cobelli, C. and G. Romanin-Jacur (1975), Structural identifiability of strongly connected biological compartmental systems, *Med. Biol. Eng.*, 831-37.

[20] Contreras, M. and G. Casella (1996), *Integral Transform Parameter Estimation*, Tech. Report BU-1363-M, Biometrics Unit, Cornell University, Ithaca, NY.

[21] Davidian, M. and D. M. Giltinan (1995), *Nonlinear Models for Repeated Measurement Data*, Chapman & Hall, London.

[22] Delforge, J. (1981), Necessary and sufficient structural condition for local identifiability of a system with linear compartments, *Math. Biosci.* **54**, 159-80.

[23] Dennis, J. E., Jr. and R. Schnabel (1983), *Numerical Methods for Unconstrained Optimization and Nonlinear Equations*, Prentice-Hall Series in Computational Mathematics, Prentice-Hall, Englewood Cliffs, NJ.

[24] Edelstein-Keshet, L. (1988), *Models in Biology*, Random House, New York.

[25] Eisen, M. (1988), *Mathematical Methods and Models in the Biological Sciences*, Prentice-Hall, Englewood Cliffs, NJ.

[26] Fletcher, R. (1987), *Practical Methods of Optimization*, 2nd ed., Wiley, New York.

[27] Fowkes, N. and J. Mahony (1994), *An Introduction Mathematical Modelling*, Wiley, New York.

[28] Fox, W. W., Jr. (1970), An exponential surplus yield model for optimizing exploited fish populations, *Trans. Am. Fish. Soc.* **99**, 80-8.

[29] Gibaldi, M. and D. Perrier (1982), *Pharmacokinetics*, Marcel Dekker, New York.

[30] Giordano, F. and M. Weir (1985), *A First Course in Mathematical Modeling*, Brooks-Cole, Pacific Grove, CA.

[31] Godfrey, K. (1983), *Compartmental Models and Their Applications*, Academic Press, New York.

[32] Groetsch, C. W. (1993), *Inverse Problems in the Mathematical Sciences*, Vieweg Braunschweig, Wiesbaden, Germany.

[33] Grosslein, M. D., R. W. Langton, and M. P. Sissenwine (1978), Recent fluctuations in Pelagic fish stocks of the Northwest Atlantic, Georges Bank region in relationship to species interactions, *International Council for the Exploration of the Sea. Symposium on the Biological Basis of the Pelegic Fish Stock Management*, No. 25.

[34] Hearon, J. Z. (1963), Theorems on linear systems, *Annals NY Acad. Sci.* **108**, 36–68.

[35] Jacquez, J. A. (1996), *Compartmental Analysis in Biology and Medicine*, Ann Arbor, MI:BioMedware.

[36] Kemeny, J. and J. Snell (1962), *Mathematical Models in the Social Sciences*, MIT Press, Cambridge, MA.

[37] Klamkin, M., Ed. (1987), *Math. Modelling: Classroom Notes in Applied Mathematics*, SIAM, Philadelphia.

[38] Leontief, W. W. (1966), *Input–Output Economics*, Oxford University Press, London.

[39] Lindstrom, M. J. and D. M. Bates (1990), Nonlinear mixed effects for repeated measures data, *Biometrics* **46**, 673-687.

[40] Littell, R. C., G. A. Milliken, W. W. Stroup and R. D. Wolfinger (1996), *SAS system for mixed models*, SAS Institute Inc., SAS Campus Drive, Gary, North Carolina.

[41] Lucas, W. F., Ed. (1983), *Modules in Applied Mathematics*, (4 volumes), Springer-Verlag, New York.

[42] Maki, D. and M. Thompson (1973), *Mathematical Models and Applications*, Prentice-Hall, Englewood Cliffs, NJ.

[43] Marcus, M. and H. Minc (1964), *A Survey of Matrix Theory and Matrix Inequalities*, Prindle, Weber and Schmidt, Boston.

[44] Maurer, R. (1975), A preliminary description of some important feeding relationships, *International Commission for the Northwest Atlantic Fishery*, Research Document 75/IX/130.

[45] May, R. M. (1973), *Stability and Complexity in Model Ecosystems*, Princeton University Press, Princeton, NJ.

[46] Meyer, W. (1984), *Concepts of Mathematical Modeling*, McGraw-Hill, New York.

[47] O'Connor, D. J. and D. M. Di Toro (1970), Photosynthesis and oxygen balance in streams, *J. Sanitary Eng. Div., SCE* **96**, 547–71.

[48] Odum, H. T. (1956), Primary production in flowing waters, *Liminol. and Oceanogr.*, 102–17.

[49] Olinick, M. (1978), *An Introduction to Mathematical Models in Social and Life Sciences*, Addison-Wesley, Reading, MA.

[50] Ore, O. (1963), *Graphs and Their Uses*, Mathematical Association of America, Washington, DC.

[51] Patten, B. C., M. C. Barber, and J. T. Finn (1976), Review and evaluation of input–output flow analysis for ecological applications, Univ. of Georgia, Contr. to Systems Ecology, #42, preprint.

[52] Pella, J. J. and P. K. Tomlinson (1969), A generalized stock production model, *Bull. Inter-Am. Trop. Tuna Commun.* **14**, 421–96.

[53] Perelson, A. S. et al. (1997), Decay characteristics of HIV-1-infected compartments during combination therapy, *Nature* **387**, 188–91.

[54] Pielow, E. C. (1969), *An Introduction to Mathematical Ecology*, Wiley, New York.

[55] Pope, J. G. and O. C. Harris (1976), The South African pilchard and anchovy stock complex, an example of the effects of biological interactions between species on management strategy, *ICNAF*, Selected Papers No. 1, 157–62.

[56] Rabenstein, A. L. (1982), *Elementary Differential Equations with Linear Algebra*, Academic Press, New York.

[57] Roberts, R. (1976), *Discrete Mathematical Models*, Prentice-Hall, Englewood Cliffs, NJ.

[58] Schaefer, M. B. (1954), Some aspects of the dynamics of population important to the management of the commercial marine fisheries, *Bull. Inter-Am. Trop. Tuna Commun.* **1**, 25–56.

[59] ———— (1968), Methods of estimating effects of fishing on fish population, *Trans. Am. Fish. Soc.* **97**, 231–41.

[60] Searle, S. R., G. Casella, and C. E. McCulloch (1992), *Variance Components*, Wiley, New York.

[61] Soong, T. T. (1971), Pharmacokinetics with uncertainties in rate constants I, II, and III the inverse problem, *Math. Biosci.* **12**, 235–243.

[62] Taillie, C. (1977), *The Mathematical Statistics of Diversity and Abundance*, Ph.D. Thesis, Penn State University, University Park, PA.

[63] Tsokos, J. O. and Tsokos G. P. (1976), Statistical modeling of pharmacokinetic systems, *Transaction of the ASME, Journal of Dynamic systems, measurement, and control*, 37-43.

[64] Venables, W. N. and B. D. Ripley (1997), *Modern applied statistics with S-Plus*, Second edition, Springer, New York, New York.

[65] Verhulst, P. F. (1838), Notice sur la loi que la population suit dans son acroissement, *Corr. Math. Phys.* **10**, 113.

[66] Walter, E. (1987), *Identifiability of Parametric Models*, Pergamon Press, New York.

[67] Walter, E. (1982), *Identifiability of State Space Models*, Lectures Notes in Biomathematics, Vol. 48, Springer-Verlag, Berlin.

[68] Walter, G. G. (1986), Size identifiability of compartmental models, *Math. Biosc.* **81**, 165–76.

[69] ———— (1984), Complexity of compartmental models, *Math. Biosci.* **70**, 147–59.

[70] ———— (1983), Passage time, resilience, and structure of compartmental models, *Math. Biosci.* **63**, 199–213.

[71] ———— (1980), Stability and structure of compartmental models of ecosystems, *Math. Biosci.* **51**, 1–10.

[72] ———— (1979), A compartmental model of a marine ecosystem, in *Compartmental Analysis of Ecosystem Models*, Matis, Patten and White, Eds., International Co-op Pub., Fairland, MD, pp. 29-42.

[73] ———— (1979), Compartmental models, digraphs, and Markov chains, in *Compartmental Analysis of Ecosystem Models*, Matis, Patten, and White, Eds., International Co-op Pub., Fairland, MD, 295–310.

[74] ———— (1976), Nonequilibrium regulation of fisheries, *ICNAF*, Selected Papers No. 1, 129–40.

[75] ———— (1975), Graphical methods for estimating parameters in simple models of fisheries, *J. Fish. Res. Board Can.* **32**, 2163–68.

[76] Wan, F. (1989), *Mathematical Models and Their Analysis*, Harper and Row, New York.

[77] Wu, H. and A. A. Ding (1998), Population HIV-1 dynamics in vivo: applicable models and inferential tools for virological data from AIDS clinical trials, *Biometrics*, **55**, No. 2, June, 1999.

[78] Zemanian, A. H. (1965), *Distribution Theory and Transform Analysis*, McGraw-Hill, New York.

Index